W9-CMO-655

Strategies Promoting Success of Two-Year College Students

ACS SYMPOSIUM SERIES **1280**

Strategies Promoting Success of Two-Year College Students

Laura J. Anna, Editor
Montgomery College
Rockville, Maryland

Thomas B. Higgins, Editor
Harold Washington College
Chicago, Illinois

Alycia Palmer, Editor
Montgomery College
Rockville, Maryland

Kalyn Shea Owens, Editor
North Seattle College
Seattle, Washington

Sponsored by the
ACS Division of Chemical Education

American Chemical Society, Washington, DC

Distributed in print by Oxford University Press

Library of Congress Cataloging-in-Publication Data

Names: Anna, Laura J.
Title: Strategies promoting success of two-year college students / Laura J. Anna, editor, Montgomery College Rockville, Maryland, Thomas B. Higgins, editor, Harold Washington College, Chicago, Illinois, Alycia Palmer, editor, Montgomery College, Rockville, Maryland, Kalyn Shea Owens, editor, North Seattle College Seattle, Washington.
Description: Washington, DC : American Chemical Society, 2018. | Series: ACS symposium series 1280 | "Sponsored by the ACS Division of Chemical Education." | Includes bibliographical references and index.
Identifiers: LCCN 2018036995 (print) | LCCN 2018055548 (ebook) | ISBN 9780841232914 (ebook) | ISBN 9780841232921 (alk. paper)
Subjects: LCSH: Chemistry--Study and teaching (Higher)--United States. | Community colleges--United States. | Junior colleges--United States.
Classification: LCC QD47 (ebook) | LCC QD47 .S74 2018 (print) | DDC 540.71/173--dc23
LC record available at https://lccn.loc.gov/2018036995

The paper used in this publication meets the minimum requirements of American National Standard for Information Sciences—Permanence of Paper for Printed Library Materials, ANSI Z39.48n1984.

Distributed in print by Oxford University Press

Foreword

The ACS Symposium Series was first published in 1974 to provide a mechanism for publishing symposia quickly in book form. The purpose of the series is to publish timely, comprehensive books developed from the ACS sponsored symposia based on current scientific research. Occasionally, books are developed from symposia sponsored by other organizations when the topic is of keen interest to the chemistry audience.

Before agreeing to publish a book, the proposed table of contents is reviewed for appropriate and comprehensive coverage and for interest to the audience. Some papers may be excluded to better focus the book; others may be added to provide comprehensiveness. When appropriate, overview or introductory chapters are added. Drafts of chapters are peer-reviewed prior to final acceptance or rejection, and manuscripts are prepared in camera-ready format.

As a rule, only original research papers and original review papers are included in the volumes. Verbatim reproductions of previous published papers are not accepted.

ACS Books Department

Contents

Undergraduate Research Opportunities

NSF Funding Programs

Indexes

Preface

Nearly half of all undergraduate students enroll in a two-year college, and many STEM students take their introductory chemistry courses at a two-year college. The educational pathway of these students is often complex and non-linear towards completion of a certificate, degree and/or transfer to a four-year program. These non-traditional pathways present distinct challenges in and out of the classroom for the educators who are trying to serve this important and diverse group of students.

This symposium series book provides some insight into the current state of chemistry education reform at two-year colleges, provides specific ways to address some of the challenges of educating two-year college students, and reports strategies for success in chemistry at the two-year college. Our contributing authors are chemical educators who have presented at one or more of our *Strategies Promoting Success of Two-year College Students* symposia that have been held at the Spring ACS National meetings since 2015.

Our authors are dedicated to the unique educational environment of the two-year college and share their innovations contributing to the success of two-year college students in certificate or degree completion, workforce development for STEM careers, and transfer to baccalaureate programs.

Overview of Strategies for Promoting Success of Two-Year College Students

The organization of this book allows readers to explore multiple strategies that chemical educators at two-year institutions use and implement at course and program levels to help guide students towards success in chemistry. Chapters are centered around themes of student engagement and collaborations, curricular innovations, undergraduate research opportunities and NSF funding information.

Student Engagement and Collaborations

The first selection of chapters shares programs and projects that support student engagement and collaborative opportunities that contribute to student success. A Learning Assistant program implemented at Montgomery College describes how a peer learning support model can introduce students to STEM teaching and positively impact student success in the classroom (Chapter 1). Synergistic efforts to engage students in the classroom through innovative inter-disciplinary content modules, early introduction to research and extracurricular supports all contribute to the support of early STEM students at North Seattle College (Chapter 2). Supplemental Instruction implemented at Bronx Community College showed effective impact on student success in chemistry and related STEM courses (Chapter 3). An NSF-funded grant project to support shared, portable instrumentation between a two-year institution and four-year institutions describes a successful collaboration resulting in positive change to expand student access to instrumentation to promote student success across all institutions (Chapter 4).

Curricular Innovations

The next section transitions to a focus on specific curricular changes at course and program levels. Contributing authors from State Fair Community College describe the development of a new pre-professional program to support students in their rural community (Chapter 5). Student affective state and its implications on chemistry course pre-requisites and instruction are examined from a student success viewpoint at East Los Angeles College (Chapter 6). The impact of incorporating in-class worksheets to support active classroom learning and improve student engagement in general chemistry courses at the University of Arkansas Rich Mountain is described (Chapter 7). At Flathead Valley Community College, a toolbox approach containing an array of effective student success strategies is described (Chapter 8).

Undergraduate Research Opportunities

Promoting student success through undergraduate research programs or experiences in the two-year chemistry curriculum is highlighted through work at three institutions. Contributing authors from Southwestern Michigan College (Chapter 9) and Montgomery College (Chapter 10) share how they collaborate with their college honors programs to recruit and motivate undergraduate students to participate in research projects. We also hear the story of the creation of an independent research program at Lorain County Community College through a collaborative partnership with a local whiskey company (Chapter 11).

NSF Funding Programs

For individuals looking to funding agencies to support expanding collaborations, curricular innovations or research opportunities, the book concludes with a chapter of tips and guidance for navigating the NSF grant proposal writing and review process (Chapter 12).

We hope that the ideas and approaches from chemical educators who have successfully integrated strategies at course and program levels to help students succeed in chemistry will be a valuable resource to others looking to implement new ideas, specific strategies or curricular innovations for student success in introductory chemistry courses and two-year college programs.

Acknowledgments

We are grateful to our contributing authors who devote time and energy every day to the success of students from the classroom through their transfer and completion goals. We are also thankful to our many anonymous peer-reviewers for their thoughtful and constructive comments that improved each contribution. We are particularly grateful for the assistance of the entire ACS editing team that made this opportunity possible so that our authors' work can be shared at a broader level across the chemical education community.

Laura J. Anna
Chemistry Department
Montgomery College
Rockville, Maryland 20850, United States

Thomas B. Higgins
Department of Physical Sciences
Harold Washington College
Chicago, Illinois 60601, United States

Alycia Palmer
Chemistry Department
Montgomery College
Rockville, Maryland 20850, United States

Kalyn Shea Owens
Department of Chemistry
North Seattle College
Seattle, Washington 98103, United States

Student Engagement and Collaborations

Chapter 1

Trying on Teaching: Transforming STEM Classrooms with a Learning Assistant Program

Carolyn P. Schick*

Learning Assistant Program Director, Department of Chemistry, Montgomery College, 51 Mannakee Street, Rockville, Maryland 20850, United States
***E-mail: Carolyn.Schick@montgomerycollege.edu.**

The nationally recognized Learning Assistant (LA) model, originating at the University of Colorado, Boulder (CU-Boulder), has been adapted for the two-year college setting at Montgomery College. LAs are recruited to assist in STEM classrooms and laboratories with a variety of unique assignments designed to enhance collaborative learning and student engagement. LAs get the opportunity to 'try on teaching' as they work alongside their faculty mentors, assisting students in the classes they support. Highlights on the faculty mentorship and reflections on teaching for the LAs are presented. Academic success, for both the students in the LA-supported classes and the LAs themselves, plus positive transformations in the STEM classrooms are discussed. The LAs, their faculty mentors, the classroom students, and the field of STEM education all benefit from this collaboration.

Introduction: The Learning Assistant

A Learning Assistant (LA) is an undergraduate student who, through the guidance of course instructors and a special pedagogy course, facilitates discussions among groups of students in a variety of classroom settings that encourage student engagement and responsibility for learning (*1*). Serving as an LA gives students a chance to 'try on teaching' for a semester under the mentorship of a faculty member. The LA gains a much deeper understanding of the material in the course and grows in confidence and interpersonal skills while working with the students in the class. The students greatly benefit by having the LA embedded in their class, where the LA serves as a peer mentor, assisting and encouraging them in a variety of ways that foster active and collaborative learning. Faculty members benefit as they reflect on and develop activities to incorporate their LAs into the classroom, enhancing the learning environment for their students.

Many STEM faculty are familiar with the TA (Teaching Assistant) model from their own graduate school experiences. A TA works to assist the *teacher*, whether it is supervising lab sections, grading exams and lab reports, or leading recitation sections for large lecture-based classes. A Learning Assistant, alternatively, is embedded in the class as an extra support to assist the *learner*. This extra support can assume many forms such as guiding students through in-class problem-solving exercises in small collaborative learning groups or on an individual basis, assisting students in the laboratory in partnership with a faculty mentor, and hosting study and review sessions for students outside of class. LAs are not permitted to act as supervisors in the laboratory setting, nor do they grade or perform other course management tasks that are typically reserved for faculty.

There are several features that distinguish the LA model from other types of learning support that might be found in a college setting, such as tutoring at a STEM learning center, Peer-Led Team Learning (*2*), or Supplemental Instruction (*3*). Learning Assistants: 1) are embedded directly into the learning environment, 2) reflect on teaching and learning pedagogy during their experience, and 3) develop a relationship with their faculty mentors throughout the semester. These features will be highlighted throughout the rest of this chapter.

The National LA Model

The national model for Learning Assistant programs began formally at the University of Colorado, Boulder (CU-Boulder) in 2003 (*4*, *5*). Supported by a National Science Foundation grant (DUE-0302134), the goal at CU-Boulder was to transform large lecture STEM classes, where they could recruit and prepare talented STEM majors for careers in teaching while at the same time target efforts to improve the quality of STEM education for all undergraduates. CU-Boulder already utilized graduate and undergraduate students as TAs in many of their STEM classes so additional undergraduate students were trained and hired as LAs. The LAs were embedded in the learning environment to increase student interaction, engage students in collaborative problem-solving, and encourage students to articulate and defend their ideas in the classroom, either in the large lecture space or in smaller discussion settings.

Within the national LA model, Learning Assistants engage in three key activities, depicted in Figure 1 (*1*).

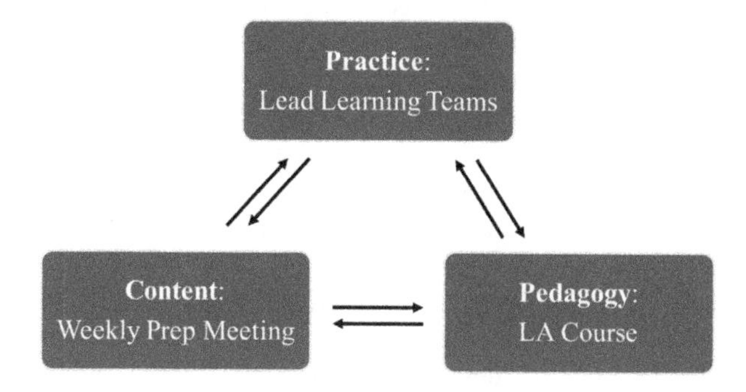

Figure 1. Learning Assistants engage in three key activities.

In "Practice", LAs interact with groups of students in the classroom environment, assisting them by facilitating discussions during collaborative learning activities and supporting their learning. Learning Assistants are themselves supported in this practice in two main ways: by attending weekly prep meetings with their faculty mentors to discuss course content and by participating in a pedagogy course concurrent with their LA assignment. In "Content", LAs and faculty mentors examine course content and discuss strategies for engagement and interaction with the students in the class. In "Pedagogy", the LA course serves to introduce LAs to education research, active learning, and concepts such as building from student ideas, listening and questioning, and building relationships with students. Through the pedagogy course, content, and their practice, the LAs are then able to integrate these educational learning ideas into effective teaching practices.

The benefits of having an LA program at an institution are numerous and multi-faceted. It has been documented that the presence of LAs in STEM classrooms contributes to student success by lowering DFW rates (students receiving a grade of D, F, or W) in gateway courses (*6*). Studies where concept inventories have been administered to classes with and without LAs have shown that LAs help students gain better and deeper understanding of the material (*7, 8*). Learning gains for students in underrepresented minority groups are greater when these students are supported by classes with an LA (*9–11*). Not only do students benefit during the semester in their LA-supported class, but positive longitudinal effects have been observed for these students in their successive classes (*12*). The LAs also benefit, as evidenced by having higher concept inventory scores in the course they are supporting in addition to having increased graduation rates and higher GPAs compared to their peers (*13*). Positive institutional change has been noted, driven by faculty mentors seeking to transform their classrooms to have more and enhanced student-centered activities due to the addition of LAs in their classrooms (*14*). Finally, teacher recruitment and the field of STEM education have benefitted as LAs come away with a positive teaching experience early in their academic careers. This leads to an increase in the number of students actively pursuing teaching as a career (*4*), and, for those LAs who ultimately become STEM teachers, they are observed to incorporate more reformed teaching practices in their classrooms (*15*).

Since 2003, LA programs have been increasing in number across the country. As of Fall 2017, there are over 80 LA programs in STEM departments in colleges and universities throughout the country (*1*). As a way to support and unify these programs as well as to continue development of the national LA model, the Learning Assistant Alliance was created, supported with grant funding and housed at CU-Boulder. The LA Alliance serves as a clearinghouse to connect LA programs worldwide and offers support to colleges and universities seeking to house LA programs at their own institutions. The LA Alliance maintains a leadership council, facilitates annual international and regional workshops across the country, hosts a growing data warehouse for LA activities, hosts online pre/post assessment tools, and provides links, documents, videos, and resources for institutions to network and gain support, either as they are starting out or sustaining and growing their own LA programs (https://www.learningassistantalliance.org/). Over 300 institutions worldwide have accounts on the LA Alliance website.

A pedagogy course, one of the general program elements of the national LA model, is typically a 1-2 credit course designed for new LAs to explore STEM teaching from a metacognitive standpoint. The LAs attend this class together, analyze texts focused on teaching and learning theory, and have opportunities to reflect on and experiment with what they are learning, either in the pedagogy class through role-playing or in the class where they serve as an LA. It should be noted that there is some additional variance in the credit level of the pedagogy course at institutions nationwide, with some institutions offering a 3-credit course, and others offering a 0-credit seminar, each driven by what makes sense in the climate of their individual institution.

As LA programs have crossed over into the two-year college setting, the implementation of the national LA model is not focused on transforming large lecture environments, as most two-year college STEM courses are already in small classrooms. At the two-year college, LA programs serve to increase student engagement and collaborative learning, improve student success rates, promote teaching as a career, and create a positive environment for mentoring. Unlike most four-year institutions with access to upper-level undergraduate and graduate students, two-year colleges do not typically employ TAs and most classes are taught and supported solely by teaching-focused faculty. Due to both transfer articulation issues and financial limitations of the LAs, a credit-based pedagogy course is often not practical to implement at the two-year college. Therefore, other strategies need to be developed to incorporate the metacognitive and pedagogical aspects of teaching into the LA experience so as to map to the national LA model. Two two-year colleges have developed pedagogy courses in a creative manner and are highlighted next.

Harold Washington College (HWC), a two-year college in downtown Chicago, Illinois and part of the City Colleges of Chicago system, has created a 16-week one-credit hybrid LA pedagogy course (*16*). The LAs at HWC do not incur additional tuition or fees associated with this course due to the college's tiered tuition structure. For the majority of the semester, the LAs participate in online discussions as they read education literature and watch video episodes. They submit reflections based on readings, teaching practice, and their own LA experience. They also work on small individual projects and have three face-to-face meetings during the semester.

The Boulder County Campus of Front Range Community College (FRCC) in Colorado, has partnered with CU-Boulder enabling FRCC LAs to participate in a CU-Boulder sponsored two-credit online pedagogy course (*17*). The FRCC LAs engage in readings, discussions, assignments, and projects in partnership with a select group of CU-Boulder LAs also taking the pedagogy course in this format. (Other CU-Boulder LAs enroll in the traditional face-to-face pedagogy course.) The online course culminates in a final video project with each of the LAs showcasing one aspect of their LA experience to their cohort LAs in the virtual environment.

Both Harold Washington College and Front Range Community College have found creative ways to offer a credit-based LA pedagogy course that fits each of their unique two-year college environments. The online format of the pedagogy course is being considered elsewhere in other partnerships between two-year colleges and four-year institutions.

The next section will describe how the national LA model has been adapted for the students and faculty at Montgomery College.

The LA Program at Montgomery College

Snapshot of Montgomery College

Montgomery College is a county-based two-year college located in the Maryland suburbs of Washington, D.C. There are approximately 24,000 students enrolled in credit coursework spread across three campuses, several off-campus locations, and online (*18*). Montgomery College has a diverse student body with a distribution that is 28% black, 24% Hispanic, 24% white and 11% Asian. Over 160 countries are represented, 76% of the students are considered underrepresented minorities, and 53% of the students are female. In Fall 2017, there were 13,000 students enrolled in STEM classes with 280 declared chemistry/biochemistry majors. The chemistry department supports students taking introductory chemistry through second semester organic chemistry as well as additional upper-level bioanalytical and biochemistry courses. Montgomery College partners with nearby National Institutes of Health (NIH), National Institute of Standards and Technology (NIST), and local biotech companies to provide students with internships and research opportunities to complement their academic studies. Students can also embark on research projects in chemistry and other STEM areas at the college, either through honors modules (*19*) or a credit-based independent research course.

Program Overview

With initial support from a National Science Foundation grant (DUE-1239965, Robert Noyce Capacity Building program), the LA Program began at Montgomery College in Fall 2013 with 16 LAs in a variety of STEM courses. Since that time, the program has grown to approximately 40 LAs in a multitude of STEM courses (preparatory, gateway, and upper-level) across the three campuses of the college, serving over 1000 students each semester. The LA program has supported 240 individual LAs from Fall 2013 through Fall 2017, with 17 of these LAs having a second assignment the following semester. Approximately 7,000 students have directly benefitted from having an LA in one or more of their STEM classes. Since inception at Montgomery College, the LAs and the program have been supported by multiple grants with a focus on student success and increased diversity in STEM (*20*). Internal college support from the three STEM Deans (Chemical and Biological Sciences, Mathematics and Statistics, and Science, Engineering, and Technology) as well as the three campus-based Vice Presidents and Provosts has been invaluable as the college actively embraces initiatives focused on increasing student success.

Typical Duties

There are a variety of ways that faculty mentors can embed an LA into their STEM classrooms to promote active learning and increased student engagement. Faculty mentors specify the blend of assistance desired in their courses based on their teaching style and course goals. Below are the types of assistance that LAs can provide at Montgomery College.

- Assistance in the classroom: LAs provide assistance in flipped classroom settings and facilitate group activities as students are working on and discussing problems in both lecture and recitation/discussion sections. LAs are instructed on the Socratic method of questioning when interacting with students and are encouraged to ask guiding questions without providing direct answers. In addition, LAs have the opportunity to give mini-lectures and present example problems on the board to the students in the class.
- Assistance in the laboratory: LAs provide an extra set of trained hands and minds to assist students with carrying out and making sense of their experiments in the laboratory setting. LAs interact with students as they are performing their experiments and help groups of students with post-lab analysis. The LAs are guided on how to approach students by asking them open-ended questions which allows for greater and deeper understanding of the experiments and underlying concepts. In addition, LAs have the opportunity to give pre-lab lectures, present demonstrations of related laboratory equipment, and perform fun demo experiments.
- Leading study and review sessions: The LAs form strong peer-mentor relationships with the students in the class as the semester progresses. Because LAs have formed a positive relationship with the students

through being embedded in the learning environment and endorsed by their faculty mentor, they are effective at encouraging students to attend weekly study sessions and periodic review sessions before quizzes and exams. Because the LAs are exposed to the teaching style of their faculty mentor by being embedded in the class, LAs design the study and review sessions in a way that complements the faculty mentor's approach. During the sessions, LAs answer student questions, lead students to develop concept maps, guide them as they work on review packets, encourage group and team work, and help the students process the course material with a big-picture perspective. LAs learn to work with students in ways that help the students gain confidence and successfully navigate key concepts. As the LAs are working with the students in these outside settings, there is also opportunity for them to share study skills, note-taking skills, and academic advice, all of which serves to promote student success.

- Online support: Although this is a minor component, LAs might also assist with online student questions in a course management environment, such as Blackboard, or in a text-messaging based group chat environment, such as GroupMe (*21*). Through Blackboard and with guidance from their faculty mentors, LAs initiate discussion threads, pose questions to students, and help support an active discussion board throughout the semester. To that end, LAs develop reflective questioning skills and learn how to effectively promote discussions between students in the online environment. Under the supervision of the faculty mentor, both the students in the class and the LA can also sign into the course's GroupMe account, set up by the faculty mentor. Students can ask the group a question and anyone, including the LA, can respond. This type of interaction works well to further strengthen the positive peer-mentor bond between LAs and the students as the LAs interact with the students in a social media environment that students often prefer. LAs can also remind students of upcoming study and review sessions, help clear up misconceptions, and answer questions in these types of class-based forums.

The LA Journey

The journey of an LA assignment begins by faculty members requesting an LA for one of their courses prior to the start of the semester. Students then apply and interview for the requested positions. LAs are hired and paid a stipend for their work as an LA. They are required to commit approximately six hours per week to their LA position. Additionally, the cohort of LAs meet three times during the semester for training and pedagogy development. Their participation in these gatherings builds a sense of community among the LAs themselves. LAs give a presentation that is formally evaluated by their faculty mentor during the semester and they are presented with an LA Certificate upon completion of the semester. Online and paper-based surveys from the various participants (LAs, students, and faculty mentors) are collected throughout the semester for program evaluation.

Figure 2 shows the main components of the Montgomery College LA program with each segment being described in detail in the following subsections.

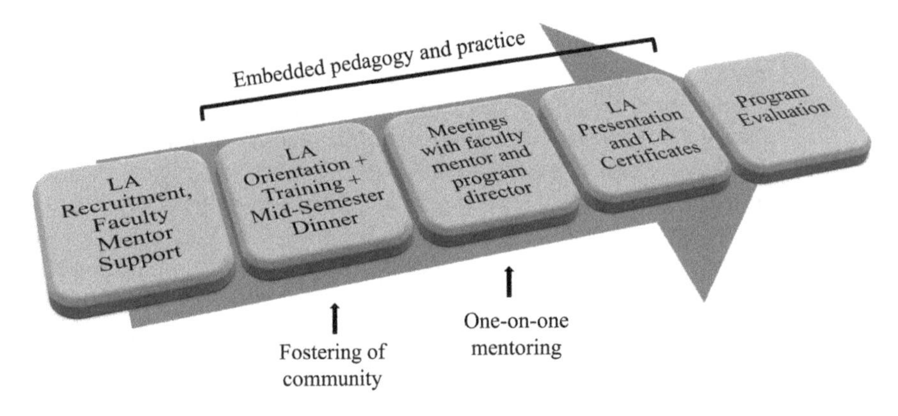

Figure 2. The semester-long journey of an LA assignment.

LA Recruitment Process

After departments have finalized their teaching assignments for the upcoming semester, faculty have the opportunity to formally request an LA for one of their STEM courses (biology, chemistry, computer science, engineering, mathematics, or physics). The faculty mentors decide how best to embed an LA in their courses, whether the LA is to be placed in the lecture, discussion section, laboratory, or a combination of settings. In their request proposals, faculty mentors are asked to describe the benefits of having an LA embedded in their learning environments, both for the students in the class and for the individual LA. To that end, faculty mentors must reflect on their own teaching style and practice and be open to developing enhanced opportunities for active student engagement and collaborative learning as they think through incorporating an LA into their classrooms. There is flexibility for adjustment as faculty members select and hire their LAs, keeping in mind the strengths and goals of their individual LA coupled with pragmatic scheduling concerns. Faculty buy-in is strong; this is completely voluntary and faculty are not compensated by the college for mentoring a student as an LA each semester.

The recruitment of Learning Assistants has two components: 1) the program director reaching out to all students who have recently taken the appropriate courses, and 2) faculty reaching out to students in their own classrooms. Personalized emails to all qualified students are sent out in addition to posting announcements through the college's student site and providing slide media to be displayed on hall video monitors across each campus. Faculty contribute to the recruitment process by making announcements in class, having their current LAs talk about the program in class during the recruitment phase, and approaching individual students who might have an interest in teaching. Both components of recruitment are essential, as it is important to the integrity and growth of the LA program that all STEM students be invited to apply, not just those who have seen an LA in action or been approached by a faculty mentor. During the selection and interview process, faculty mentors are encouraged to resist selecting only their personally known candidates. Once they work with their first LA who was previously unknown to them, faculty mentors have anecdotally shared that any bias they might have had for selecting only their known students has been dissolved.

After students apply for the advertised LA positions, the applicants are screened for appropriate prerequisites, overall academic stability, scheduling availability, and approval by faculty references. Each faculty mentor is then given several candidates to interview. The LA interview is a vital component of the program because it ensures that the selected LA can meet the goals of the individual faculty mentors. During interviews, faculty mentors observe how each applicant might explain a problem to a confused student and they also analyze the approachability, maturity, and work ethic of the applicants to ensure the selected student is a good fit for their course. Given that this program is voluntary for faculty, it is crucial that the faculty be actively involved in the selection and approval of the LA who will ultimately be a significant part of their classroom environment.

Candidates who are interviewed, but not ultimately chosen, still gain a positive experience. The interviews are framed as a growth opportunity for all students versus a dead-end exercise if a candidate is not selected. Both faculty mentors and the LA Program Director often follow up with the remaining candidates, offering encouragement and appropriate academic advisement. Support during and after the interviews is often another grounding and launching point for these students along their academic path. Many of the candidates who are not selected with their first application, apply the following semester. With more experience and development, these students are often accepted on their second try and are excited to finally join the LA program.

Faculty Mentor Support

Prior to the start of each semester, the LA faculty mentors gather for a working lunch meeting to learn about program development and share ideas with each other on how to effectively work with their LAs in their classrooms. New mentors are welcomed and have the opportunity to discuss program specifics with their peers. Veteran faculty mentors continue to be inspired and exchange ideas during this meeting, creating a living, healthy program where suggestions are shared and faculty understand their role in national LA model. Conversations between faculty mentors include discussions on how to guide LAs to interact effectively with student groups during class, strategies for increasing participation and attendance at LA-led study sessions, and types of presentations the LA can give during the semester. A future growth area in the faculty mentor meeting could be adding a professional development component centered around a pedagogical reading where the faculty mentors would then discuss the reading as a group, and explore how they, their classrooms, their LAs, and their students can all benefit. The faculty would then be poised to incorporate their new ideas into their LA-supported classrooms as the semester begins.

LA Orientation, Training, and Mid-Semester Dinner

At the beginning of each semester, the LAs come together as a group for two separate gatherings a few weeks apart – orientation and training, led by the LA Program Director. LAs get to meet and interact with each other during these events. This is especially important because, for the majority of the semester, the LAs are on their own working with their faculty mentors in an embedded class. For the LAs, understanding where they fit in the college and in the LA program creates a sense of belonging and growing enthusiasm. It also inspires them to hold themselves with greater responsibility as successful student peer models not only for the Learning Assistant program but in their other classes at the college as well.

At the two-hour LA Orientation, professionalism and conduct as a student employee/LA are discussed as well as issues associated with student privacy. The LAs introduce themselves to each other, learn about the national LA model, and learn about the LA program at Montgomery College. They view video clips of other LAs in action at the college so they can visualize themselves growing in this program. The framework of the LA experience is laid out with initial thoughts on teaching gathered and discussed.

At the six-hour LA Training, LAs learn about and discuss a range of topics and teaching strategies, such as ways to encourage independent student learning, appropriate expectations for both LAs and students, the Socratic method of questioning, dealing with challenging or underprepared students, open-ended vs. closed-ended questions, valuing and eliciting student ideas, fostering a collaborative learning environment, giving students tools to solve their own problems, and finding available resources to refer to students. Exploration and development of these ideas and strategies in LA Training is key, as this is one of the ways pedagogy is integrated into the LA experience at Montgomery College. The LA Training is run jointly between the LA Program Director and the STEM Learning Center Director, who hires and works with STEM tutors each semester. Both directors attend and contribute to the two training events (LA training and tutor training). The partnership with some shared overlapping content gives scheduling options to both LAs and STEM tutors if they are not able to attend the training with their original group.

One worry that LAs often express is how to handle student questions when they don't know the answer. At training, the LAs practice how to say, "I don't know, let's look through your notes for the answer" or "That's a really good question, let's figure it out together". This type of approach has two advantages: 1) it encourages students to be independent learners, not relying on a quick answer provided by someone else without first activating their own critical thinking skills, and 2) LAs have the opportunity to model how they would search for an answer, which is an important academic skill to develop.

To prepare the LAs for when student groups are working together on a problem or worksheet assignment in class, they also receive guidance on how to encourage direct dialog between the group members, how to ensure all voices are heard, and how to help groups work toward convergence as the members exchange ideas and listen to each other. The LAs reflect on the importance of being a facilitator of group discussion, recognizing that their primary role is that of an outside guide or observer. This is an important skill to practice and develop as the LAs learn how resist their own instincts to directly answer a question to one member of a group which could then degrade the collaborative learning environment that the groupwork was designed to promote in the first place.

Sometimes during training, LAs take part in a communication challenge. In pairs, one LA verbally describes an object, pictured on an index card, to the partner using solely positional and geometric words. The partner LA then tries to create the object using modeling clay. At the end of the challenge, the pair compares the resulting molded object with the original image on the card. This fun exercise gives the LAs practice at communicating effectively and demonstrates the importance of good verbal and listening skills between teacher and learner.

The LAs also work in small groups to create and perform skits which address how to handle different situations that might come up in their work and how to lead students to be successful learners. Each of the LAs has a role in these skits as they all gain practice with their presentation skills in front of an audience. The skits typically have a fun element to them ('here's what not to do', or 'here's how an LA works with a difficult student' - where one LA pretends to be the difficult student). The LAs enjoy this component of training where they laugh together as they try out different techniques to use in their own LA assignments. The LAs also get to practice approaching students, facilitating group interactions, and waiting in silence to give thinking a time to mature, all of which are skills that they will use throughout their experience embedded in the classroom.

At the three-hour LA Mid-Semester Dinner event, all of the LAs are invited to gather together again. In addition to providing LAs with an opportunity to share their experiences as a cohort, this event features a pedagogy-based presentation, organized and led by the LA Program Director. While presentation themes vary each semester, some representative topics include: Culturally Responsive Teaching (*22*), Failure isn't Fatal (*23*, *24*), Growth vs. Fixed Mindsets (*25*), and Metacognition in Teaching (studying versus learning, Bloom's taxonomy, The Study Cycle, etc.) (*26*). Prior to attending this event, the LAs complete a survey that asks them to reflect on their teaching perspectives and other topics relevant to the presentation. This survey prepares them to discuss these topics productively in small groups at the event. One of the goals of the mid-semester dinner is to give the LAs more tools and reflective points that they can then pass on to the students in the class. The LAs are encouraged to make changes in their own lives in response to what they learned with their fellow LAs, with the positive ripple effect starting with themselves, and moving outward to the students in the class as well as to their faculty mentors. LAs are encouraged to report back to their faculty mentors about the mid-semester dinner, sharing what they gained from the event. The process of explaining can serve to further cement new perspectives and strategies the LAs have learned at the dinner.

Mentor Meetings

Throughout the semester, the LAs develop a supportive relationship with their faculty mentors. Each pair is required to meet weekly to discuss content and plans for the upcoming week, along with approaches to address common misconceptions and ways to interact effectively with students in the class. Pedagogy is integrated into these weekly meetings as the pair discuss different teaching strategies and techniques to use in the class and with the students. The weekly meeting is also useful for the faculty mentors to gain student perspective through the lens of the LA, and gauge where students in the class might be stuck or in need of more clarification. Having the faculty mentors discuss student perspective with their LAs is invaluable and allows for any mid-course corrections that might be necessary.

Because the LA and faculty mentor are in this supportive setting, it is natural for them to discuss the LA's own academic and career goals throughout the semester. The faculty mentor serves as an informal academic advisor in this capacity and provides guidance on such topics as choosing a major, navigating a career path, selecting a transfer institution, or even exploring internship opportunities. At the start of the semester, both the LA and the faculty mentor discuss goals they have for the experience and these weekly meetings are a good opportunity to check in on these goals. For an LA goal, an LA might want to improve his/her public speaking skills, so the pair could use these meetings to discuss ways to reach this objective. For a faculty mentor goal, faculty mentors might want to have the LA facilitate new activities for increasing student interaction in class, so the pair could use these meetings to reflect on the effectiveness of these activities for learning success. Both the LAs and the faculty mentors can find their voice through this relationship, which can have far-reaching benefits, such as cultivating an appreciation and interest for teaching as well as fostering a growth mindset for the LAs along their academic journeys.

About halfway through the semester, the LA Program Director goes to each campus to meet individually with the LAs for a check-in meeting. Pragmatically arranging these in-person meetings is often challenging due to the busy commuter-based schedules of the LAs and the multi-campus environment of the college, however, these meetings serve an important purpose for supporting the LAs and ensuring that the program is running smoothly. The LA's workload is checked to make sure the faculty mentor's expectations are appropriate and that the six-hour weekly commitment is being honored. Because this meeting is not focused on content, the pair discuss the big picture and any issues the LA might be having such as strategies for managing a group of students in a study session, the LA's own academic progress during the semester, dynamics of working with their faculty mentor, time management, and moments of triumph as an LA. These meetings between the LA and the Program Director have a holistic objective – focusing on the success of the LA as a student, a learning assistant, and a person, which is especially important in the two-year college environment where students are often juggling life, work, family, finances, and school simultaneously. Oftentimes there is follow-up for a particular issue, which then results in the LA being even more successful in their position moving forward.

LA Presentation and LA Certificates

Each LA is required to give at least one presentation/mini-lesson in the class which is formally evaluated by the faculty mentor. Most LAs give more than one presentation during the semester. These presentations range from giving a pre-lab lecture, to working a few problems on the board in the classroom or discussion section, to giving a brief presentation on a topic supported by the faculty mentor and part of the class curriculum. The faculty mentor prepares written feedback on an evaluation form and then meets with the LA to discuss the presentation, giving both compliments and suggestions for improvement. The LAs receive a copy of this formal evaluation and can incorporate it into their future teaching portfolios. This enables the LAs to look back on their experience with concrete evidence of what might have been their first moments in front of a classroom.

After the semester ends, the LAs receive a Certificate of Completion. LAs are invited to serve a second semester, although new applicants are given priority. If an LA is hired for a second semester, he or she is hired to work with a different faculty mentor, supporting a different course, preferably in a different STEM discipline. This new assignment serves two purposes: 1) the veteran LAs increase their knowledge and expertise in another course, which benefits their academic career, and 2) they get to work with a different faculty mentor, giving them increased exposure to diverse teaching styles and new experiences as they are embedded in another classroom to 'try on teaching' again.

Program Evaluation

Throughout the semester, the LA program is assessed through completion of online and paper-based surveys. Both the faculty and LA surveys are a required part of the program, therefore almost 100% compliance is achieved every semester, resulting in a complete assessment of the program by the LAs and faculty mentors. All of the surveys are analyzed and programmatic changes are made to better serve the LAs, the students, the faculty mentors, the LA program, and the college as a whole.

- **LA Pre-Survey** – This online survey is completed by the LAs before the semester begins. The survey assesses interest in teaching and asks the LAs to reflect on what they hope to gain through the experience. The LAs are asked if they have previously taken an LA-supported class, if they know any friends who are LAs, what their academic major and career interests are, to describe any prior teaching/tutoring/mentoring experience, and if/how they knew their faculty mentor prior to their LA position.
- **LA Mid-Semester Reflections** – This online survey is completed by the LAs before the mid-semester dinner event. The survey asks LAs to describe one of the best teachers they've had and what makes that person a good teacher, what qualities of a best teacher they hope to bring to their LA experience, their most rewarding session so far, and what they would like to do over if they could. This survey also asks reflective questions that prepare the LAs for the theme of the mid-semester dinner with topics that will be explored in depth at the event.

- **Student Evaluation of LAs** – At the end of the semester, faculty mentors are given the opportunity to have their LAs evaluated by the students in the class. This is not a mandatory component of the program as it is important for faculty to choose how to use class time especially at the end of the semester. Typically over 65% of the faculty mentors incorporate the student evaluations as part of their course experience, with almost all of the students in their class participating. Using either paper forms distributed in class or an online questionnaire, the students are asked to evaluate the experience of having an LA in their class. They evaluate the LA and articulate how working with the LA affected their learning and success in the course. The students are also asked to describe the LA's strengths and offer suggestions for improvement. As evidenced by their thoughtful responses, students are eager to give earnest appreciation and feedback to their LA, who they often view as being an invaluable asset in helping them successfully navigate the course. At the end of the semester, the LAs receive copies of their student evaluations for self-reflection of their journey as an LA.
- **LA Post-Survey** – This online survey is completed by the LAs at the end of the semester. The survey again assesses interest in teaching, and asks for reflection on how the LA experience has influenced their interest in teaching, what the LAs gained from the experience, whether or not they felt supported by their faculty mentor, what was helpful to them, and suggestions for the program as a whole.
- **Faculty Mentor Post-Survey** – This online survey is completed by the faculty mentors at the end of the semester. The faculty mentors reflect on how having an LA benefitted both the students in the class as well as themselves with their own teaching development, they analyze the one-on-one weekly meetings, they share a favorite moment working with their LA, and they have the opportunity to give programmatic suggestions for the future.

Positive Impacts of the LA Program at Montgomery College

Impact on Students: Academic Success and Attitude

Although not presented here, embedding an LA into a STEM gateway course has had a positive impact by lowering DFW rates when comparing STEM sections with and without LA support, both at Montgomery College and around the country (*6*).

As mentioned earlier, LAs are evaluated by the students in the class at the end of the semester. These evaluations help gauge the LA's impact on students' attitudes towards learning. Results from 387 Fall 2017 student evaluations are presented in Table 1 with the percentage listed representing agreement with each of the corresponding statements.

Table 1. Student Attitudes Towards Learning

Statement	*Agreement*
The LA encouraged me to think.	98.4%
The LA encouraged me to participate and ask questions.	95.8%
The LA answered my questions effectively.	96.9%
The LA has increased my appreciation for the course material.	94.5%
I have learned more by having an LA in this course.	92.2%

Figure 3 shows the expanded positive response for the statement "The LA encouraged me to think." The data from these qualitative assessments on learning shows that the students in the class highly value the added assistance of an LA for their own learning gains and success in the class.

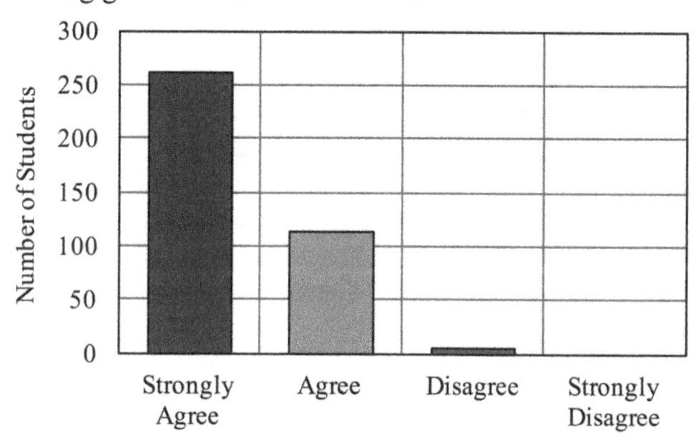

Figure 3. Students in LA-supported classes responded to the statement, "The LA encouraged me to think."

When students are asked how having an LA in the class helped them, they often share thoughts of improved academic performance as shown in several quotes from student evaluations:

> *"My learning experience was greatly enriched because of the LA, and I don't think I would've understood the material that well without his assistance."*

> *"I appreciate the after-class review sessions the LA would hold, it has definitely helped me improve my test grades."*

> *"The LA has been encouraging. He has a way of answering your questions by asking you questions as a guide towards the actual answer. I found myself just talking out loud and telling him my thought process, and that in itself was enough for me, so just having him there to listen and agree or disagree with me was a lot of help."*

These students directly connect the LA's support and pedagogically minded guidance to their own increased performance in the class.

Students also share how the LA nurtured positivity and confidence as shown in the following student evaluations:

"Most beneficial, the LA has helped me understand the material when I was confused and boosted my confidence."

"When I felt intimidated asking the teacher questions on anything in regard to the course contents and materials, the LA made me feel comfortable asking any questions I wanted."

"I was more confident because the LA helped me by supporting my thoughts on certain topics."

"Having an LA has given me the comfort to learn."

By exposing the LAs to the growth mindset concept during the training and mid-semester dinner discussions, LAs are better equipped to pass this mindset on to students, offer encouragement, and make them feel comfortable with the idea that persistence and positive thinking can often lead to academic success.

One design feature that helps LAs connect with students is that they are students themselves. The following student evaluation quotes demonstrate the power of peer learning support:

"The LA was able to get along with all students by connecting with us and explaining the course material from a student perspective."

"It's good to learn from someone who is pretty much a student like me."

"The LA was close to my age, understood the struggle and motivated me to be better."

Having the LA be a relatable peer provides enhanced trust in the learning relationship.

Students describe how advantageous it is to have a "second teacher" able to work with them, both in and outside of class. The following student evaluation quotes demonstrate the extra resource the LA provides:

"Whenever the professor was busy helping other students, the LA could help us while we were struggling. The LA helped us a lot." .

"During Discussion, when we are working on our problems, the professor cannot help every person. The LA is additional help and provides different perspectives to learning chemistry." .

A few faculty mentors have shared humorous moments when students raise their hands with a question, the professor comes over and the students respond that they were actually hoping the LA could come over for assistance instead. In these instances, the LAs are respected and valued almost as equally as the professor.

Enhanced and longitudinal learning also happens for LAs. As highlighted in Bloom's taxonomy, by teaching concepts to others, LAs develop a deeper understanding of the course material compared with when they initially learned the material as a student. One example of this longitudinal benefit occurred when a student was hired as an LA for Organic Chemistry I after just having earned an average grade in the course the previous semester. As she served as an LA for Organic Chemistry I, she was concurrently enrolled in Organic Chemistry II. The LA was excited for the opportunity and the faculty mentor was glad to work with her, recognizing that there would be occasions for continued learning because she hadn't yet fully mastered all of the content. By having this LA experience, the LA was able to deepen her understanding for the Organic Chemistry I material as she prepared to work with the students in class. For this LA, there was no propagation of struggle or cumulative penalty in her sequential course, as there might have been without this experience. Her increased mastery of Organic Chemistry I material undoubtedly helped her in her Organic Chemistry II class, and, in the end, she arrived at a more full and complete understanding of the two-term sequence because of her LA experience.

Impact on LAs: Interest in Teaching

LAs are asked to state their interest in teaching before (pre-survey) and after the LA experience (post-survey). By the end of each semester, almost all LAs report an increased interest in teaching as shown in Figure 4.

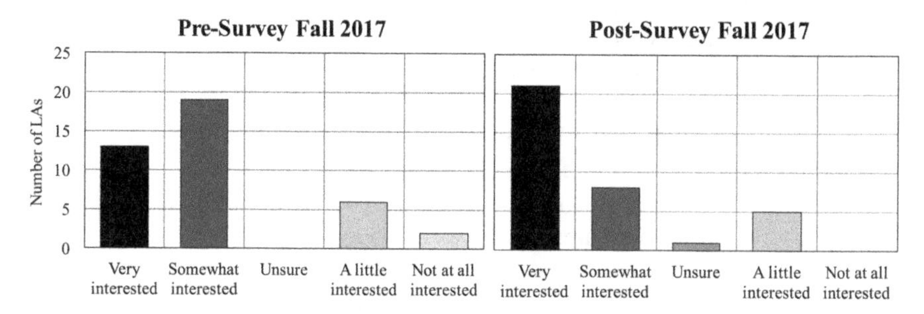

Figure 4. Before and after the semester LAs are asked, "How interested are you in a career that involves some level of teaching (i.e. high school teacher, college professor, leader/trainer/educator) in your STEM field?"

Although not seen with the Fall 2017 cohort, there have been a couple of cases where an individual LA, while having an overall positive LA experience, has reported a decrease in teaching interest after the experience concludes. This result can also be viewed with a successful mindset. These LAs were not sure about teaching and thought it might be a possible interest. Having the LAs 'try on teaching' for the semester helped them figure out that this career path was not the right fit for them. This is especially valuable for the students to realize early on in their academic journeys, so that they can gravitate toward different opportunities in the future. If this type of "self-discovery" occurs later in their academic studies, then students might waste precious time and energy focusing on something that did not suit their talents and inclinations. It is therefore of great value to give interested students an early taste of STEM teaching, to see if that interest can cultivate and grow or if their focus should be directed elsewhere.

Since Fall 2013, four LAs have received full scholarships to teaching programs at nearby four-year institutions. These students were positively affected by their LA experience and subsequently sought out, applied for, and were awarded teaching scholarships, which then gave their transfer path a STEM teaching focus. Even though the LA program has a high focus on growing collaborative learning environments in the classroom, increasing academic success for affected students, and promoting positive development in LAs, gains are seen for increasing the pool of teachers in the STEM field.

While the teaching focus of the LA program is designed to develop the teaching interest for the LAs, students in the class are indirectly affected in a positive manner. On the student evaluation forms at the end of the Fall 2017 semester, 68.0% of students in the LA-supported classes agreed with the following statement, "I have considered the possibility of teaching by watching the LA in my classroom."One student wrote on his evaluation form "It's great to see others interested in teaching."The LA program clearly cultivates positive teaching interest and enhanced learning experiences for both students and LAs throughout the semester.

Impact on Faculty: Collaborative and Active Learning Environments

For faculty who participate in the LA program, their classes can be transformed in subtle ways. As shared through informal discussion and faculty post-surveys, some faculty don't initially choose to be an LA faculty mentor with the primary objective of transforming their classes, but, as they ponder how to incorporate their newly hired LA, they end up developing more activities and opportunities for enhanced student engagement and collaborative learning in their classrooms. The incorporation of an LA into the classroom can serve to motivate faculty mentors to further embrace and foster an active learning environment in their teaching. Thus, by having LAs, classroom teaching can be transformed to be more student-centered, interactive, and supportive in an effort to reach the great diversity of learners in the classroom.

Opportunities and Challenges at the Two-Year College

Opportunities: Peer-Modeling

One component that an LA program can bring to increasing student success in the two-year college classroom is having successful students themselves serve as models to their peers. At Mongomery College, LAs are high-achieving and hard-working students who have done well in the class they support, earning an A or B in that class from a recent semester. Sixty percent of the LAs in Fall 2017 had a GPA of 3.6 or higher, with 98% of them having a GPA of 3.1 or higher. These students serve as excellent role models for the students they support in STEM classrooms.

While it is imperative that LAs be competent in the course they are hired for, LAs do not need to be content experts – that is the job of the professor. As a result, many faculty mentors tend to gravitate toward enthusiastic students willing to work hard, but who might not necessarily be at the top of every class, versus 4.0 students who might never have struggled because the course material always came easy for them. Faculty mentors come to see the advantage of choosing an LA who is relatable in struggle to the students in the class. The students in the class clearly benefit from the extra assistance and encouragement that this type of student offers.

LAs are often passionate about the subject matter and course content where they offer support. They performed strongly in this class themselves and like the material. To have an LA display genuine enthusiasm for the material in a class of students, some of whom are unsure about the material, is infectious and a powerful example for the students in the class. On student evaluation forms, students often express that the LA "made learning fun" or the LA "made me appreciate chemistry."

It is not surprising that typically half of the current LAs have previously taken an LA-supported class, even if it was in a different STEM class than the class they are currently serving. Having a wide range of STEM courses for students to choose from is an asset to the students interested in this type of experience. This allows students to select a course where they feel comfortable being an LA while also allowing them to witness and learn from LAs in other STEM courses prior to their own LA experience. This could be described as the domino effect where students taking a class with an LA are inspired to become LAs themselves, who then inspire another group of students, and the cycle continues.

Having students who are underrepresented minorities (URMs) serve as LAs is also effective for increasing student success. Eighty percent of the LAs at Montgomery College are URMs, matching the percentage of URMs at the college in STEM. This is not the result of any special targeted recruiting, but just by having a diverse student body of students excited about leading and supporting their peers. Fifty percent of the LAs are female, which also matches the gender breakdown for students in STEM at Montgomery College. Even though the make-up of LAs matches both ethnicity and gender ratios at the college, it is important to state how valuable and impactful it is that the "norm" is a successful underrepresented minority or female student serving as a peer leader in the field of STEM.

Opportunities: Mentorship

Throughout each semester, faculty form positive relationships with their LAs. Forty percent of the LAs have taken a class with their faculty mentor prior to joining the program and the rest are forging a new relationship as their LA position starts. In their LA post-survey, LAs share that they felt greatly supported and encouraged by their faculty mentors. Even at times when LAs feel they could have done better during a particular moment, faculty mentors offer concrete advice, positivity, and support. One such example was when an LA gave her mini-presentation to the class, which involved working through and explaining several problems on the board. The LA made a simple, but propagating error while writing on the board in front of the class. At the appropriate time, the faculty mentor intervened to prevent student confusion, but did so in a way that was supportive to the LA, acknowledging that everyone makes mistakes. The faculty mentor set a tone of support for the LA, and didn't leave her floundering in front of the student audience.

Faculty mentors can also discuss time management and course/academic advising with their LAs once a positive and honest relationship has been established. The faculty mentors observe organization and time management in action and are often more positioned to provide advice on future wise choices with their LAs compared to regular academic advisors who might not have had the opportunity to observe the student in this capacity. The faculty mentors can therefore capitalize on opportunities for improvement. As part of the faculty mentor post-survey, one faculty mentor wrote, "Some of my favorite moments with my LA were the long discussions we had that ranged from career paths, family, her adjustment process to living in the U.S., and social issues." Clearly these mentor relationships are encouraging and affirming to both the LAs and their faculty mentors. In the faculty mentor post-survey, another faculty mentor wrote, "I look forward to mentoring my LA every semester. It's incredible to see the confidence of the LA grow over the course of the semester, and such an advantage for students in courses with high DFW rates to have a role-model to interact with."

Because LAs are supported by their mentors in this environment, they flourish. Multiple LAs have presented their experiences in both poster and oral formats at national conferences (Robert Noyce Northeast Conference and the ACS National Meeting, Chemical Education division). It can be a life-changing moment for a student at a two-year college to present his or her experiences and perspectives as an LA on the national stage. This leadership role opens many doors for a successful academic career as LAs continue their studies in STEM. After her presentation at the Noyce Northeast Conference one LA wrote to the LA Program Director, "Thank you for your guidance. Your encouragement to do what I thought I couldn't do [present at a national conference], and the chances you gave me, have made the greatest impact. Thank you, because I couldn't have done it without your support!"

Mentorship also occurs between former LAs and current LAs. After transferring to four-year institutions, several LAs have been invited back to Montgomery College to give presentations to the current cohort of LAs. During these presentations, the LA alumni share their transfer journey and discuss how the LA experience helped propel them to where they are now. These first-hand success stories serve to motivate the current LAs to gain as much as they can from the experience. The LA alumni serve as informal mentors in this capacity as they showcase their own pathway to success.

Other Opportunities

At two-year colleges, there are more pre-college and lower-level courses for LAs to support than at most four-year colleges and universities. Many of these pre-college level classes resemble similar ones found in high school teaching environments. STEM Education students can then actively seek out these positions for their LA experience, giving them early exposure and practice that fits with their career interests.

The focus on teaching at the two-year college provides a natural environment for teaching exploration and development. Two-year college faculty now have a vehicle in which to talk to students about a potential teaching career. Without the LA program, a STEM faculty member might observe a student in the class and think 'I bet s/he would make a good teacher' but not be able to offer the student any nurturing experience to accompany the observation. Even faculty members who don't directly participate in the LA program routinely refer students to the program. It is interesting to note that referrals to the LA program go in both directions. Sometimes students, familiar with the LA program from taking another LA-supported STEM class, have professors with which they have had a positive classroom experience. These students will then approach their professors asking if there might be opportunities in their classes for an LA. Prompted by these eager students, the faculty members might then investigate and join the LA program, motivated by the prospect of supporting and mentoring students in this environment.

Students who serve as LAs at a two-year college can be recommended for or seek out similar opportunities at their transfer institutions (either with an LA program or similar peer-mentoring or academic success program). This unique transfer pathway can help to stabilize and connect students to a supportive cohort at their new school during the difficult transition phase of transfer.

Sometimes students stay an extra semester at a two-year college to complete one or two more requirements before transferring. In this case, serving as an LA for any course (lower- to upper-level) is a great opportunity for these students. As LAs, they strengthen their own knowledge and remain active, engaged, and invested in their academics as they wrap up their time at the two-year college. This experience will undoubtedly be an asset to them as they develop leadership, confidence, as well as communication and enhanced learning skills – all keys to success along their journey.

Challenges

The many benefits of the LA program do not come without challenges. Due to transfer and financial issues, the typical credit-based pedagogy course offered for LAs at four-year institutions might need to be adapted in the two-year college setting. Currently, the solution at Montgomery College has been to embed the pedagogy into the LA experience throughout the semester through the one-on-one mentorship with faculty mentors and the program director, plus training, the mid-semester dinner event, formal presentation feedback, and online reflections.

At two-year colleges, there is often a smaller recruitment pool for LA positions in upper-level courses as students taking these courses transition to a four-year institution soon after completing these courses. As a result, it can be challenging to find LAs for classes such as Organic Chemistry II, Human Anatomy II, Differential Equations, etc. However, targeted recruitment efforts by the LA Program Director and faculty members can overcome this issue and make it possible to find qualified students for these upper-level courses.

How To Start at Your Institution

For readers interested in starting an LA program at their own two-year college, the author suggests starting with a small group of enthusiastic STEM faculty and growing the LA program from there. Be mindful of diversity – approach faculty who have different classroom and teaching styles and be sure to have balanced representation from all STEM disciplines. At the annual International LA Conferences (*1*), other program leaders have shared that if they start an LA program in one department at their school, it is often very challenging to expand into other STEM departments. At Montgomery College, there have been great advantages for growing slowly, but steadily in the first few semesters – this allowed for flexibility in identifying and successfully navigating key and unique parameters at the college. It was not necessary to have all the features developed at the start of the program. With each semester on this multi-year journey, the LA program has thoughtfully added and developed new components, being mindful of the national LA model, as well as transitioning from grant funding to growing internal college support. Small changes over time will build a quality LA program. With time, creativity, and flexibility, STEM students, LAs, faculty mentors, and the college can all benefit from this student-centered and collaborative learning focused program.

Conclusion

The Learning Assistant program at Montgomery College has been an important contributor to student success for students in LA-supported STEM classes, both for LAs, and for STEM students as a whole. The LAs serve as peer models in the classroom, giving students extra support, and leading them to a deeper understanding of the material. The LAs reflect on teaching and learning theories as they interact with and encourage students to articulate and defend their ideas in the classroom. The LAs form positive relationships with their faculty mentors, helping them to grow and mature along their academic paths. STEM classes are transformed into more student-centered active learning and collaborative environments, increasing the potential for student success. LAs get to 'try on teaching' for a semester to see if they want to pursue teaching in their future. Having an LA program at a two-year college has been a win-win-win for all involved.

Acknowledgments

The author would like to express appreciation to Professor Debra Poese, Director of Teacher Education Partnerships at Montgomery College. Deb was the Principal Investigator on the two NSF Noyce grants and initially suggested the idea of incorporating Learning Assistants at the college. She was a key partner as the LA program was being developed and continues to support the program.

Many administrators have supported the growth and inclusion of the LA program at the college. The STEM Deans as well as the Vice Presidents and Provosts have supported the program in numerous ways.

In the five years of the program, many different grants have graciously supported the LA program, all having a focus on student success. In addition to the two Noyce grants, the NSF GT-STEP grant supported the program from 2014-2017. The AAC&U TIDES grant supported the program from 2014-2017.

The LA Alliance as well as numerous colleagues at CU-Boulder and other institutions have provided great resources and developmental support for the LA program at Montgomery College. The annual International LA Conference, held at CU-Boulder, has been a key opportunity to brainstorm with outside colleagues on many issues as Montgomery College's LA program has developed and grown.

The LA program at Montgomery College would not exist without the enthusiastic participation of the faculty mentors. It has been a pleasure to work with the faculty mentors, as STEM colleagues, in search of talented and energetic LAs to support their students and help transform their classes. The faculty mentors have opened their classroom doors to the LAs providing them with an incredible and unique experience at the college. The program is indebted to them for their mentorship to the students and trust in the program.

Lastly, the Learning Assistants themselves are to be heartily acknowledged. The LAs work hard and their positive attitude of encouragement toward students in the class they serve is inspiring to watch. The LAs are the ripple effect that help other students, faculty mentors, the LA program, and the college all succeed, each in their own way. One LA wrote a well-known phrase in her final post-survey, "I learned to teach, and I taught to learn." It has been a pleasure to work with and watch the LAs grow.

References

1. *The Learning Assistant Alliance, the LA Model and General Program Elements of an LA Program.* https://learningassistantalliance.org/ (accessed Dec. 30, 2017).
2. Gafney, L.; Varma-Nelson, P. *Peer-Led Team Learning: Evaluation, Dissemination, and Institutionalization of a College Level Initiative*; Springer, Netherlands, 2008.
3. Dawson, P.; van der Meer, J.; Skalicky, J.; Cowley, K. On the Effectiveness of Supplemental Instruction: A Systematic Review of Supplemental Instruction and Peer-Assisted Study Sessions Literature Between 2001 and 2010. *Rev. Educ. Res.* **2014**, *84*, 609–639.

4. Otero, V.; Finkelstein, N.; McCray, R.; Pollock, S. Who Is Responsible for Preparing Science Teachers? *Science* **2006**, *313*, 445–446.
5. Otero, V.; Pollock, S.; Finkelstein, N. A Physics Department's Role in Preparing Physics Teachers: The Colorado Learning Assistant Model. *Am. J. Phys.* **2010**, *78*, 1218–1224.
6. Alzen, J. L.; Langdon, L.; Otero, V. K. The Learning Assistant Model and DFW Rates in Introductory Physics Courses. In *2017 Physics Education Research Conference Proceedings*; Ding, L.; Traxler, A.; Cao, Y., Eds.; AIP Press: Melville, NY, 2018; pp 36−39.
7. Talbot, R. M.; Hartley, L. M.; Marzetta, K.; Wee, B. S. Transforming Undergraduate Science Education With Learning Assistants: Student Satisfaction in Large-Enrollment Courses. *J. Coll. Sci. Teach.* **2015**, *44*, 28–34.
8. Goertzen, R. M.; Brewe, E.; Kramer, L. H.; Wells, L.; Jones, D. Moving toward Change: Institutionalizing Reform through Implementation of the Learning Assistant Model and Open Source Tutorials. *Phys. Rev. Spec. Top. Phys. Educ. Res.* **2011**, *7*, 020105.
9. Van Dusen, B.; Nissen, J. M. Serving Marginalized Physics Students: an HLM Investigation of Collaborative Learning Environments. *J. Res. Sci. Teach.* **2018**(under review).
10. Van Dusen, B.; Nissen, J. M. Systemic Inequalities in Introductory Physics Courses: the Impacts of Learning Assistants. In *2017 Physics Education Research Conference Proceedings*; Ding, L.; Traxler, A.; Cao, Y., Eds.; AIP Press: Melville, NY, 2018; pp 400−403.
11. Sellami, N.; Shaked, S.; Laski, F. A.; Eagan, K. M.; Sanders, E. R. Implementation of a Learning Assistant Program Improves Student Performance on Higher-Order Assessments. *Cell Biol. Educ. Life Sci. Educ.* **2017**, *16*, ar62.
12. Pollock, S. J. A longitudinal study of the impact of curriculum on conceptual understanding in E&M. In *2007 Physics Education Research Conference Proceedings*; Hsu, L.; Henderson, C.; McCullough, L., Eds.; AIP Press: Melville, NY, 2007; Vol. 951, pp 172−175.
13. Otero, V. K. Nationally scaled model for leveraging course transformation with physics teacher preparation. In *Recruiting and Educating Future Physics Teachers: Case Studies and Effective Practices*; Sandifer, C., Brewe, E., Eds.; American Physical Society: College Park, MD, 2015; pp 107−116.
14. Pollock, S. J.; Finkelstein, N. Impacts of Curricular Change: Implications from 8 Years of Data in Introductory Physics. In *2012 Physics Education Research Conference Proceedings*; Engelhardt, P. V.; Churukian, A. D.; Rebello, N. S., Eds.; AIP Press: Melville, NY, 2013; Vol. 1513, pp 310–313.
15. Gray, K. E.; Webb, D. C.; Otero, V. K. Effects of the Learning Assistant Model on Teacher Practice. *Phys. Rev. Phys. Educ. Res.* **2016**, *12*, 020126.
16. Escuadro, A.; Langdon, L. *LA Pedagogy Contexts: Online and Hybrid.* Session presented at 2017 International Learning Assistant Conference, Boulder, CO, Nov. 4−6, 2017.
17. Gomez, C.; Langdon, L. *An LA Program Partnership to Increase Two-Year College Student Pathways into Math, Science, and Engineering.* Poster

presented at: 2017 International Learning Assistant Conference, Boulder, CO, Nov. 4–6, 2017.

18. *Montgomery College at a Glance*, 2017. Montgomery College. http://cms.montgomerycollege.edu/WorkArea/DAsset.aspx?id=63282 (accessed Dec. 30, 2017).
19. Palmer, A. M.; Anna, L. J. Honors Modules to Infuse Research into the Chemistry Curriculum. In *Strategies Promoting Success of Two-Year College Students*; Anna, L. J.; Higgins, T.; Palmer, A. M.; Owens, K.; ACS Symposium Series 1280; American Chemical Society: Washington, DC, 2018.
20. Support from NSF Robert Noyce Teacher Scholarship Program (Capacity Building), Award No. 1239965 (2012-2015); *NSF Graduate and Transfer STEM Talent Expansion Program*, Award No. 1161231 (2012-2017); AAC&U Teaching to Increase Diversity and Equity in STEM, (2014-2017); and NSF Robert Noyce Teacher Scholarship Program (Phase I Scholarships and Stipends), Award No. 1555634 (2016-2021).
21. *GroupMe*. https://groupme.com/ (accessed Dec. 30, 2017).
22. Gay, G. *Culturally Responsive Teaching: Theory, Research, and Practice*, 2nd ed.; Teachers College Press: New York, 2010.
23. *The Resilience Project*. http://learningconnection.stanford.edu/resilience-project (accessed Dec. 30, 2017).
24. Sobel, A. How Failure in the Classroom Is More Instructive Than Success. *Chron. High. Educ.* **2014**. https://www.chronicle.com/article/How-Failure-in-the-Classroom/146377 (accessed Dec. 30, 2017).
25. Dweck, C. S. *Mindset: The New Psychology of Success*; Ballantine Books: New York, 2008.
26. McGuire, S. Y. *Teach Students How to Learn: Strategies You Can Incorporate Into Any Course to Improve Student Metacognition, Study Skills, and Motivation*; Stylus Publishing: Sterling, VA, 2015.

Chapter 2

Synergistic Efforts To Support Early STEM Students

Kalyn S. Owens*,[1] and Ann J. Murkowski[2]

[1]Chemistry Department, North Seattle College, 9600 College Way N., Seattle, WashingtonA 98103, United States
[2]Biology Department, North Seattle College, 9600 College Way N., Seattle, Washington 98103, United States
***E-mail: kalyn.owens@seattlecolleges.edu.**

Two-year colleges have a unique opportunity to broaden participation in STEM, but in order to do this, they must be innovative and willing to drive change. North Seattle College (NSC) has been implementing reform around three primary areas – innovative curriculum, early research, and extracurricular supports – to synergistically support a diverse array of early STEM students. These efforts all focus on the need to move beyond the mastery of discipline-specific content and towards a vision of a holistic pathway that nurtures students as learners, scientists, and engaged members of a supportive, STEM-focused community. Essential to this approach is the development of educational experiences that establish a "culture of thinking" where student voices are encouraged, valued, and driven by engagement with relevant interdisciplinary and investigative activities.

Introduction

In his 2011 State of the Union address, President Barack Obama declared that we were facing "our generation's Sputnik moment" to spur innovation in science. In 2012, the President's Council of Advisors on Science and Technology (PCAST) released their powerful report, *Engaged to Excel,* detailing a range of best practices and recommendations for educating a STEM workforce ready to meet this challenge (*1*). These recommendations come at an urgent moment when more than half of college students who declare an interest in STEM as freshman leave these fields before graduation (*2*). At our two-year colleges, the attrition is even more dire; 70% of these STEM-identified students will change fields (*3*). The PCAST was just one of many leading agencies calling for dramatic reform in the way we design STEM programs, courses, and curriculum (*4–6*). This reform is crucial to both meeting projected workforce demands in STEM and fostering the 20th century thinking skills future STEM professionals will need to solve complex problems. In addition, it has the potential to broaden participation in STEM, improving the quality and diversity of our workforce while providing access to rewarding careers for minoritized students.

North Seattle College (NSC), a member of the Seattle College District, has been developing a multi-faceted approach to support early STEM students. NSC is an urban campus that enrolls approximately 9000 students annually from diverse backgrounds and is federally designated as both an AANAPISI (Asian American Native American Pacific Islander Serving Institution) eligible institution and as a Minority Serving Institution. Like most two-year colleges, NSC struggles with declining state funding, variable enrollments, an over-reliance on adjunct faculty, and a wide range of student preparation. Despite these challenges, we recognize the unique opportunity of the two-year colleges to dramatically affect change by leveraging their flexibility and access to the most diverse population of potential STEM students during their most critical, early STEM experiences. Collectively, two-year colleges enroll 45% of all undergraduates, including 52% of black, 57% of Hispanic, 43% of Asian/Pacific Islander, and 62% of Native American students (*7*). Within these numbers lies an opportunity for community colleges to substantially impact the trajectory of this diverse population.

To harness this opportunity, the vision for two-year college STEM programs must go beyond a focus on mastery of discipline-specific content in isolated foundational STEM courses. The priority must be placed on the holistic development of individual students as learners, scientists, and highly functional members of a community as described in the PCAST report, Figure 1 (*1*).

Importantly, the PCAST report identifies the need to intellectually and personally engage students as the first two essential elements of a successful STEM program. While there are many approaches to engaging students, we highlight here how three primary efforts at NSC function synergistically to address the PCAST recommendations in a holistic, student-centered approach: 1) providing evidence-based innovative curriculum, 2) creating early research experiences focused on student development, and 3) sustaining an ecosystem of extracurricular supports (Figure 2).

Elements of Successful STEM Education Programs
(PCAST, 2012)

Essential Element	Approach
Intellectually Engage Students	• Use evidence-based methods to engage students in **creating and integrating knowledge** • Involve students in **research early**, preferably as freshman • **Build alliances** between 2- and 4-year universities to enhance research experiences • Facilitate structures that enable **group learning**
Personally Engage Students	• **Show relevance** of STEM subjects to human and planetary problems • Provide **role models** with diverse backgrounds • Promote **STEM communities** in classes, research labs, and extracurricular activities • Build **classroom communities** in which students feel they are being groomed for STEM fields rather than weeded out • Accommodate the needs of **non-traditional students**
Educate Faculty	• Provide faculty with **training in teaching** • Provide graduate students and postdocs with training in teaching • Provide faculty with **collections of learning tools** and technology
Assess Outcomes	• **Assess understanding through diverse means** and articulate assessment with learning goals • Evaluate teaching in terms of learning goals and how they are assessed and met • Assess student retention in major

Figure 1. Elements of Successful STEM Education Programs (1). The report identifies four key areas of support including the need to intellectually and personally engage students, educate faculty, and assess outcomes.

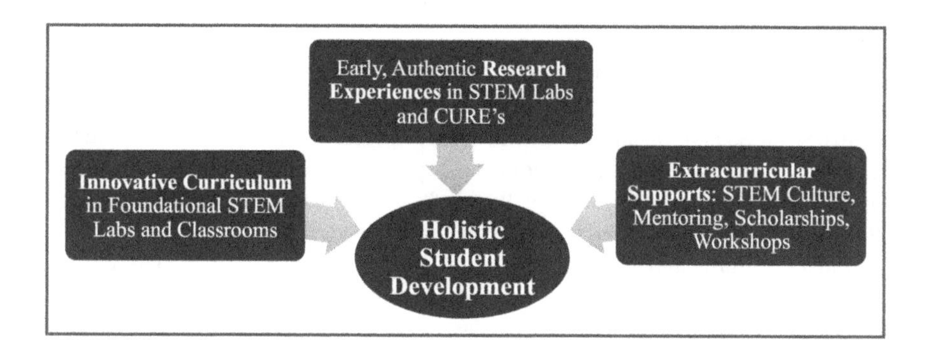

Figure 2. North Seattle College Model for Development of STEM Identity. NSC has pursued three critical essential components to best support a diverse array of developing STEM students.

Innovative Curriculum

Traditional STEM courses are generally focused on delivering discipline-specific content knowledge, to the exclusion of using these early courses as an opportunity to lay the foundation for developing students into complex thinkers, communicators, and collaborators. The national effort to transform these courses is predominantly focused on design of curriculum and pedagogical strategies that promote active learning of conceptual science (*8*). Active engagement has been shown to be an effective strategy for achieving content related outcomes, but is perhaps just the tip of the iceberg when it comes to what is needed to address the national crisis in STEM education. Innovative curriculum is needed that goes beyond active engagement towards a classroom environment designed for social construction of new knowledge to the point of deep understanding of interdisciplinary threshold concepts, while also working towards the development of students into complex thinkers, communicators, and collaborators.

Interdisciplinary Investigations (IDI) for the Classroom

To address the need for innovative curriculum in early STEM courses, our initial focus was on the design and implementation of curriculum for the general chemistry series that is interdisciplinary, investigative, and centered on the application of threshold concepts in a classroom setting. The aim of this work is to shift the focus of early STEM courses towards a curriculum that combines the benefits of an engaging collaborative experience with the powerful learning and development opportunities that emerge from weaving disciplines together and exploring open-ended questions. The curriculum has roots in a long-term collaboration between the chemistry and biology departments at NSC and has evolved over time into a set of interdisciplinary modules that can be adopted by any general chemistry program. These modules are called interdisciplinary investigations (IDIs) and are designed for both the classroom (IDI-Classroom) and for the lab (IDI-Lab).

IDIs are designed around relevant issues, as we recognize the need to motivate, build confidence, and address the ongoing problem of retention of students in STEM disciplines. Substantial research illustrates that students often feel that the STEM disciplinary topics they encounter in classes are inaccessible and irrelevant to their lives (*9*). This is especially true for students from non-dominant cultures who may have met fewer scientists than those from dominant cultures and who hold different value systems (*10*, *11*). The current IDI-Classroom modules for the general chemistry series are: 1) *Epigenetics: Investigating the Structure and Function of DNA*, 2) *Aquaporin: Investigating the Structure and Function of Cell Membrane Water Channel Proteins, and 3) Hemoglobin: Connecting Transition Metal Chemistry and Thermodynamics to Oxygen Transport*. Each module contains a short video and reading material to capture the interest of early science students, such as the recent discovery of mammoth blood in a frozen specimen, allowing for the comparison of mammoth and elephant hemoglobin in the third IDI.

Design-based research methodology was used as a means to iteratively design and redesign each IDI, which is based on the collection and analysis of student work and student observations while participating in IDIs. Essential to the design was Mansilla's assertion that interdisciplinary learning must first be grounded in a strong understanding of the individual disciplines (*12*). Students learn chemistry and biology concepts first and then engage in problem solving that requires both disciplines to solve. Ultimately this process resulted in an IDI design framework that is based on three key actions: 1) *connect students to prior knowledge, 2) extend their thinking into other disciplines and the scientific literature* and 3) *challenge participants through addressing a complex problem that requires multiple disciplines to understand* (Figure 3).

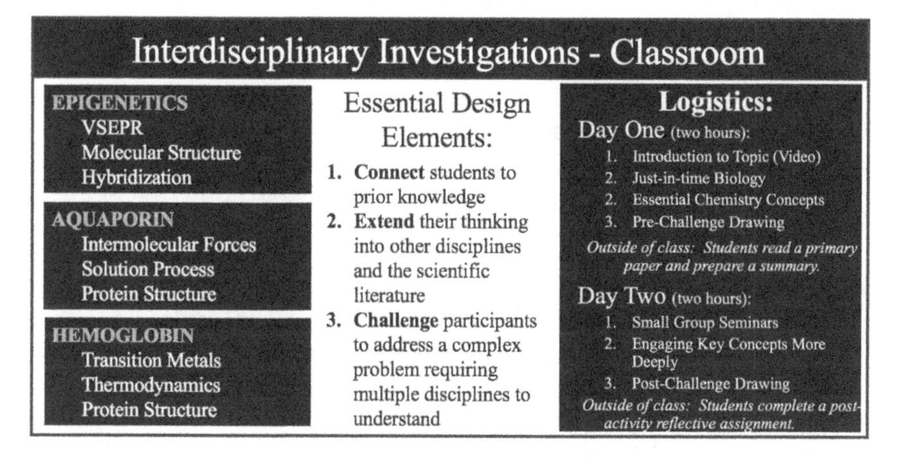

Figure 3. Interdisciplinary Investigations for General Chemistry (IDI-Classroom). Three IDIs were developed for the general chemistry series using three essential design elements: Connect, Extend, Challenge. IDIs can be completed in four classroom hours, which equates to two class days at NSC.

The Aquaporin IDI, for example, connects the concepts of polarity, intermolecular forces (IMFs), solution process and protein structure, and centers on a relevant topic of interest—the Nobel Prize for work on this fascinating protein and new aquaporin mimics used for water purification. The IDI extends to relevant biology content that has been identified as significant for understanding the module. In this case, cell membrane structure including the fluid mosaic model and membrane bound proteins are introduced, along with osmosis. The interdisciplinary challenge involves using both disciplines to create a model for the aquaporin water channel and how water moves through the channel. Students work in groups to draw and ultimately present a model that represents the thinking of the group. During the second session, students engage in a seminar (structured small group discussion of the scientific literature), learn more about the chemistry and biology of the topic, and once again create a model of the aquaporin water channel. Students then complete a post-activity assignment that involves reflection on the creative and interdisciplinary work they completed in class.

Connecting each IDI to a biochemical *threshold* concept (*13*) is a significant goal of this work. The Aquaporin IDI is designed to connect most directly to intermolecular forces (IMFs) and solution process topics in the general chemistry content, both of which have been identified as problematic for students (*14*), as well as foundational for moving on to more advanced courses (*13*). To take this on, the Aquaporin IDI activity involves substantial in-class drawing time of all the IMFs associated with cell membrane structure as well as chemical explanations for the alignment of molecules in cell membranes. Protein structure and the forces involved with how a membrane bound protein sits in the membrane are also investigated with significant amount of time spent on identifying these forces and drawing them. Groups then create a drawing on a tabletop white board of water moving through a water channel in and out of a cell with the forces clearly illustrated. While we are interested in several outcomes that result from student participation in the Aquaporin IDI, we are particularly interested in documenting student progress on their conceptual understanding of intermolecular forces in the context of this complex system especially in their ability to draw hydrogen bonds correctly. We are in the process of carrying out a study that compares students who participate in IDI to those that do not in terms of the ability to draw hydrogen bonds correctly (*15*).

While *what* is taught is clearly significant to most educators, it is the *how* that is perhaps in most need of transformation when it comes to better supporting early STEM students. With this in mind, IDIs have been developed to incorporate an array of pedagogical strategies that promote collaboration, deep thinking, visible learning and metacognition (Figure 4). This approach not only provides a mechanism to address outcomes that are often left out of early STEM courses, but also provides an opportunity to work with faculty on employing new approaches in the classroom. Making learning visible through in-class white board drawing is at the heart of a pedagogy that fuels our ability to achieve an expanded set of outcomes. Drawing to learn engages students in complex thinking, collaboration, and application of unsettled knowledge to new contexts. This approach also allows for "assessment at a glance" and through documentation of student drawings, provides an opportunity to engage students in metacognitive activities that result in long term reflection on the learning process and on self as a learner.

Understanding the impact of the IDI-Classroom approach on student development is a major focus of our current work. Three questions form the foundation of our inquiry: 1) How can we document in-process interdisciplinary learning at the interface between chemistry and biology as it occurs in a classroom setting? 2) What qualities and thinking processes are present in a meaningful interdisciplinary learning episode in general chemistry courses? 3) Do IDI participants retain a deeper understanding of threshold concepts compared to non-IDI students?

Pedagogical Strategies in Interdisciplinary Investigations	
Pedagogy	**Strategy**
Provocation	• Provoke interest through relevant problems and issues using engaging videos and journal articles
Socially-Mediated Construction of Knowledge	• Construct knowledge in small groups • Analyze and discuss primary journal articles in a structured group seminar activity • Restructure classroom environment to promote active collaboration
Representation	• Use small group white board drawing activities to represent thinking • Use molecular model kits in combination with molecular level drawing • Create model to explain interdisciplinary process using group drawing activity
Documentation and Metacognition	• Project student drawings to the class for discussion and reflection • Capture student thinking using video and show it to class for on-going reflection • Use follow-up assignment that engages students in reflection about their own thinking

Figure 4. Pedagogical Strategies Embedded in IDI-Classroom Modules. IDI's are designed around four critical, evidence-based, pedagogical strategies. This design facilitates student learning and models classroom strategies for faculty.

To begin our research efforts and to document student progress during an IDI, 88 students were paired and in-class drawings were collected (22 drawings on Day 1 and 22 drawings on Day 2). Video was used to capture student thinking during the Day 2 drawing session. Day 1 and Day 2 drawings were coded using a modified version of Mansilla's assessment framework for analyzing interdisciplinary thinking (*16*). Drawings were analyzed for purposefulness, disciplinary grounding, integration, and creativity. Students progressed in all four categories with similar gains in the first three categories (Figure 5).

Collecting and analyzing student drawings provides a unique window into the thinking that occurs in small groups as students apply new concepts, integrate ideas from two disciplines, and build a model to explain a complex system. Capturing dialogue using video provides another tool for analyzing interdisciplinary thinking; we are currently working towards using this video to effectively document and apply language to the types of gains students achieve as a result of engaging in an IDI. Ultimately we seek to articulate the value of the IDI-Classroom approach as innovative curriculum that expands the outcomes we can achieve in early STEM courses such as the general chemistry series.

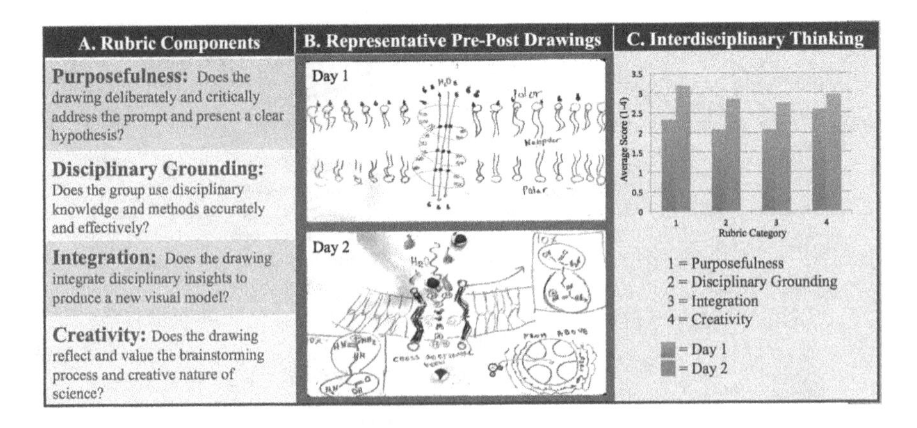

Figure 5. Using Student Drawings to Document Interdisciplinary Thinking. A. A modified version of Mansilla's interdisciplinary thinking assessment framework (16) was used to score student drawings in four categories: purposefulness, disciplinary grounding, integration and creativity. B. Day 1 and Day 2 small group drawings were collected from thee different sections with three different instructors. One example of a Day 1 and Day 2 drawing is shown here. C. The drawings were coded on a scale of 1-4 (1 = not present, 2 = novice, 3 = apprentice, 4 = mastery) and reported as an average for Day 1 and Day 2.

Early Research

Combined with innovative curriculum for the classroom, engaging students in early research experiences as a part of the academic journey provides an additional opportunity to transform the first two years of STEM programming into a more holistic and developmental approach (*17*). As illustrated by several studies, undergraduate research experiences provide an opportunity to immerse students in a unique combination of activities as a means to achieve diverse cognitive and behavioral outcomes (*18–20*). Participation in research also advances student learning (*21*, *22*), increases the likelihood of earning a degree, and retains diverse students in fields in which they are traditionally underrepresented (*23*). Course-based undergraduate research experiences (CUREs) are high-impact strategies that expose more students to the benefits of participating in authentic scientific practice (*6*, *24*, *25*), and are able to accomplish this early in the post-secondary curriculum (*26*). While there is far more research documenting the impacts of traditional UREs, it is clear that CUREs provide an exciting way forward when it comes to working with a diverse population of students, in large numbers, early in the college journey, and at institutions that do not have a culture of individual faculty running high profile scientific research programs (*6*).

Interdisciplinary Investigations (IDI) for the Lab

At NSC, we have developed a set of IDIs that are specific for the general chemistry lab program and are working to develop IDIs for the biology series. This initial set of IDIs are based on a long-term effort to implement early research opportunities (*27*). The IDI-Lab curriculum provides a method to re-envision early STEM lab courses by integrating research experiences to engage students in thinking and skill development activities that are in line with authentic scientific practices. The approach presents the opportunity for first year-college students to generate interdisciplinary research questions and work in a team to complete a small-scale project within their foundational courses. The context of each IDI involves a current and relevant issue that is rooted in the scientific literature. The curriculum is developed around a single, central, sophisticated instrument, allowing two-year colleges to more effectively leverage their limited resources.

The IDI-Lab curriculum includes three distinct phases: 1) introduction to instrumentation and the scientific literature, 2) developing a research question, and 3) a multi-week research project. The progression of phases is structured such that the research becomes more complex across a year-long series of courses and is designed to achieve selected learning outcomes at each stage with emphasis on developing investigative thinking skills. Currently the IDI curriculum complements existing lab activities, but does not replace many of the accepted approaches to meeting the core general chemistry lab outcomes. Similar to IDI-Classroom, a design framework has been created to support transfer of the model to other institutions and research contexts (*28*).

At NSC we are currently implementing the IDI-Lab approach in the general chemistry series using ion chromatography (IC) as the central research grade instrument. A parallel early research experience is also under development in the biology series using a research-grade instrument to measure photosynthesis, the Li-COR 6800, as the centerpiece of the curriculum. In both cases, the research project grows in complexity and length across the year with the focus on iteratively designing research questions. Each IDI, including both classroom and lab curriculum, involves reading one or more primary scientific journal articles, and engaging in a structured, in-class seminar. The seminar is part of a progression of activities completed as a way to expose students to the primary scientific literature in early STEM courses (Figure 6). The final stage of the year-long progression of IDI-Lab activities involves a six-week capstone research project during which time students are asked to generate a research question and have the opportunity to collect and analyze samples several times. The results of student projects are presented in the form of a scientific poster at a campus-wide symposium.

Building Fluency and Confidence with Primary Literature in Early STEM Students

Learning Outcome	Teaching Strategies	Assessment
Distinguish primary and secondary sources	• Students read a primary and secondary source; prepare summary before class • Use guided small-group discussion to elicit list of differences and value of each	Final Assessment: Literature Review Evaluated with a Detailed Rubric
Implement Effective Strategies to Read Primary Papers Independently	• Students read faculty-selected, simple primary paper; prepare summary before class • Small group guided-inquiry activity to discover the structure and function of each section • Groups report out in facilitated large group discussion	
Use Databases to Find Relevant Primary Papers	• Guided inquiry activity with librarian to discover online databases, develop effective search strategies, and filter results with Boolean operators	
Actively Participate in Effective Small Group Seminar	• Students read primary paper; prepare summary • Large group discussion on value and outcomes of seminar • Students placed in teams of four with clearly articulated roles for each member • Teams given specific questions to address from the paper • Large group reflection on both paper on seminar process	

Figure 6. Building Fluency and Confidence with Primary Literature in Early STEM Students. Deliberately designed assignments address key learning outcomes to build students' confidence and skills. These same outcomes are addressed in both the IDI-Lab and the second-year URE, allowing students to iteratively improve their skills.

Second-Year Undergraduate Research Program

A pathway from first-year CUREs (IDI-Labs) to a second-year undergraduate research program has been developed at NSC. The year-long program consists of a series of separate, for-credit courses that are designed to be student-centered, collaborative, and outcomes focused. The design of the program is specifically tuned to ensure student cognitive and skill development over time. Similar to the IDI-Lab progression in the general chemistry series, our early research experiences are focused on student growth and are open to everyone who meets the requirements for general chemistry entrance.

This year-long early URE at NSC is designed to progress through two phases with an emphasis on achievement of specific outcomes (*28*). Phase one focuses on building individual STEM identity, which is accomplished through a series of structured classroom activities, a small-scale guided research experience, and immersion into STEM culture both on and off campus. In phase two, the emphasis shifts to collaborative project development and building communication skills including writing, presenting, and collaborating in small groups. The final portion of the program (still part of phase 2) supports project completion, reflection, and strategies for success in the next steps of the academic journey including transferring to a four-year university, obtaining additional research opportunities, networking, and building a resume for professional and graduate school entrance.

The emphasis of the year-long program is student growth, development, and preparation for future STEM success. Outcomes such as using scientific articles to support project development, working effectively in small groups, and delivering effective oral presentations are taught and assessed iteratively throughout the program. Student work is evaluated on detailed rubrics that rank their skills from novice to mastery. For example, curriculum has been developed that specifically teaches small group communication and collaboration skills throughout the year (Figure 7). Students progress through a series of activities from an initial team-building activity to establishing group norms and finally to activities that involve reading and discussing the literature on building effective teams in STEM.

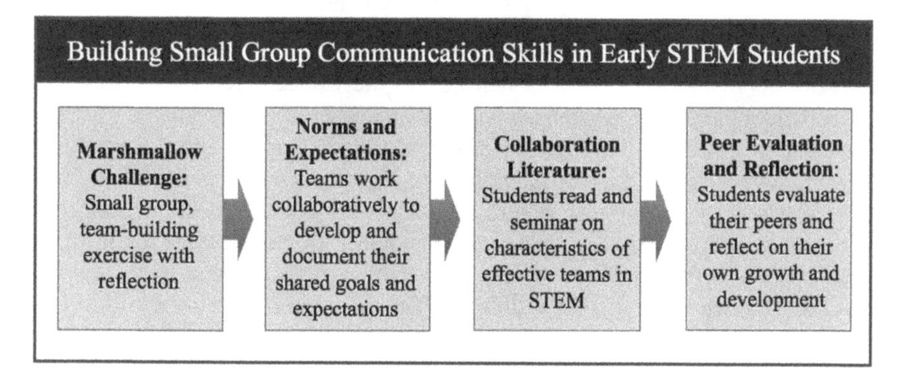

Figure 7. Small Group Communication and Collaboration Skill Development. A series of four assignments are used to develop students' interpersonal skills. Students are encouraged to reflect on their progress across the year.

Students were asked to reflect on the impact the year-long experience had on them on several different topics. When asked about their ability to collaborate and communicate in a small research group, it was clear that the URE activities impacted their awareness of the importance of collaboration. It was also clear that, in general, students were able to have a better view of themselves as a group member in terms of both their strengths and weaknesses. The following student responses are representative of their reflection on their progress in the ability to collaborate in a small group:

> Student 1: *When I get to the University of Washington, I want to push myself to seek out study groups and personal connections. Before this class, I wasn't sure what I could offer but now I feel like I have the skills to not be a drain on the group. I enjoyed working with my group and am very grateful for their help. I can see that this is how a professional network begins, and its construction seems less intimidating when you think of it a group of friends/colleagues that share mutual respect.*

> Student 2: *Within the group collaboration I was initially afraid to step into a lot of the decision making, preferring to be delegated a task, but near the end it became apparent to make a lot of the deadlines, taking initiative and completing an area or assignment was imperative for us to stay on track...... When placed in a leadership role I find myself held accountable which makes me work a bit harder, and being able to slide into this role more willingly would be a good area for improvement.*

Student 3: *I could improve my ability to discuss doubts and differing opinions. I shied away from conflict and perhaps my collaborative skills could have been improved more by taking on adversity. I did improve my ability to compromise. And I learned how to listen better to multiple opinions and to appropriately voice my own.*

A Unique REU Program for Community College Students

This outcomes focused model, as described above, was adapted to serve a broader population of students when the Seattle Colleges received a Research Experiences for Undergraduates (REU) grant from the National Science Foundation in 2014. This unique program pulled students from sixteen local community colleges, providing skills-focused research experiences in marine sciences to early career science students. Unlike traditional REU programs, the program was a yearlong experience with monthly cohort meetings to facilitate participation of place-bound students with complex lives. The program was run from and by two-year colleges to focus on the unique needs of these diverse, early career STEM students.

Program design was based on the second-year undergraduate research program at NSC, with a similar emphasis on achieving specific outcomes in a supportive learning environment. An external partner, the Ocean Inquiry Project, provided disciplinary expertise and equipment. In this model, students worked in teams to generate and pursue their own, authentic research question. While their results contribute to a deeper understanding of the Puget Sound environment, the larger goal was building resilience and skills in this population of future STEM professionals. Despite their complex lives, 92% of the REU students were able to successfully complete the year-long program and participate in a prestigious undergraduate research symposium. These students report substantial gains in their skills and understanding of the process of science (Figure 8).

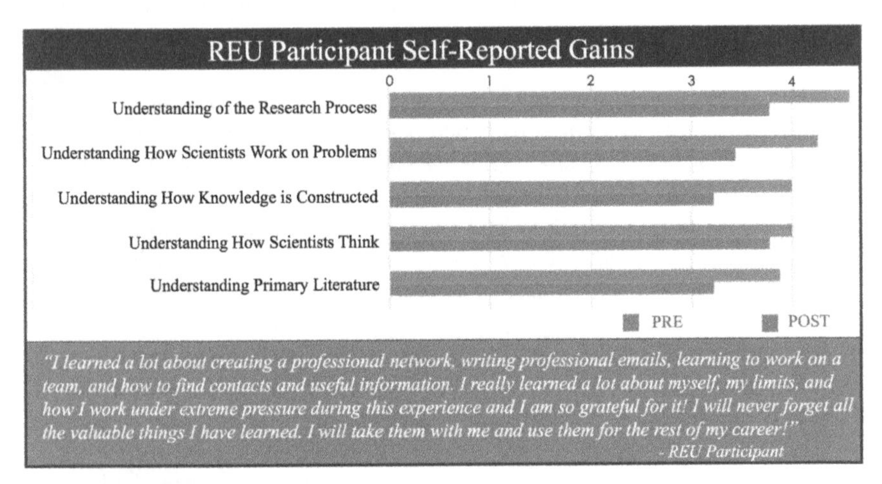

Figure 8. REU Participant Self-Reported Gains. Students report gains across numerous aspects of authentic scientific practice.

Extracurricular Supports

The third component at NSC that combines synergistically with our efforts to implement innovative curriculum and early research is the development of an ecosystem of extracurricular supports for STEM students (Figure 9). Research shows that the creation of campus-based micro-communities of peer-to-peer relationships and faculty mentors support minoritized students' long-term success in STEM (*29, 30*). These relationships foster the development of a STEM identity ultimately allowing students to see themselves as able to make meaningful contributions to their disciplines. NSC has been working collaboratively with other institutions to formally develop strategies for creating and sustaining these types of significant relationships for almost a decade. The emphasis is on increasing recruitment and retention of minoritized students in STEM through a variety of documented, high-impact practices such as faculty mentoring, cohort building, and resources directed towards student research opportunities.

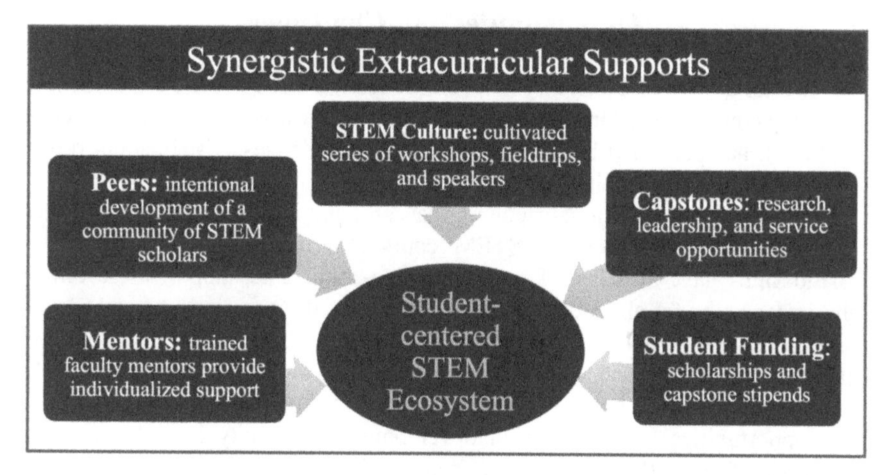

Figure 9. Synergistic Extracurricular Supports. A variety of extracurricular supports including support from faculty and peers, a strong STEM culture, the opportunity to do meaningful capstones, and student funding complement the innovative curriculum and early research opportunities at NSC.

Our formal development of extracurricular supports for STEM students began in 2009 with the formal launch of an NSF-funded program, titled Ready Set Transfer (RST). RST is a collaborative program across the Seattle College District to support students majoring in STEM fields. It provides a wide range of evidence-based supports including faculty mentoring, skill workshops, fieldtrips, and guest speakers. RST also supported student capstone experiences including undergraduate research, service learning, and student leadership. The program has served over 1000 students, including 53% first generation and 26% underrepresented minorities. RST supports these students by providing students with a faculty mentor, advising and transfer workshops, a STEM speaker series of research and industry representatives, and stipends for students and faculty mentors to collaborate in capstone experiences. RST performs well with respect to yearly retention, completion, and transfer. Compared against a matched cohort of district-wide STEM students, RST students demonstrate consistent increases in retention and completion: they had a yearly persistence rate of 92% as compared to 72%, a completion rate of 40% as compared to 22%, and a transfer rate of 37% as compared to 32% in the matched cohort.

The benefits of this suite of supports is even more dramatic when combined with scholarships that allow students to reduce the hours they work and focus on their academic and professional development. NSC has been able to graduate 95% of scholarship recipients. Across the district-wide program, 75% of these students transferred to 4-year programs, compared to just 37% of RST students and 32% of STEM students overall at the three institutions. These results demonstrate the powerful impact that scholarships combined with services have on community college students facing barriers to success in STEM fields.

Opportunities and Challenges

Working towards *excellence in STEM* education through the development of our three components (innovative curriculum, early research, and extracurricular supports), has presented both opportunities and challenges. Establishing robust collaborations with faculty at research universities, for example, provided the opportunity to co-design curriculum that brought a current and established scientific research context to STEM courses at a two-year college. Our relationships have also been key to obtaining funding to support these efforts and invigorating for all students and faculty involved. On-going interaction with research faculty provides the campus community with an infusion of current knowledge and expertise, and opens up clear pathways for next steps in STEM for students who are anxiously working their way towards the transfer point.

Opportunities for professional development for faculty have been at the forefront of our effort in working synergistically on the three components highlighted in this chapter. Two-year colleges have limited resources such that most faculty are not able to attend national conferences, nor do they generally engage in activities that require additional institutional support. Two-year colleges in Washington state have been systematically defunded over the past ten years. External funding, mostly from the National Science Foundation, has provided some of what is needed to engage faculty more deeply in the transformation of STEM courses and programming. We have used this opportunity to explore different approaches to working with faculty on innovative curriculum, early research and extracurricular supports. We provide faculty workshops around evidence-based methods for engaging students with a focus on the IDI approach in both the classroom and the lab. We model the pedagogies embedded in the IDIs in the classrooms of STEM faculty interested in the approach. We engage faculty as mentors for STEM students across the STEM disciplines and are working towards developing a best practice for working with students as faculty mentors.

Engaging adjunct faculty in this work has presented an opportunity to have an impact that goes beyond our institution. All faculty teaching in the general chemistry series at NSC have attended faculty workshops and had the opportunity to adopt the IDI curriculum in their classroom and lab program. For example, one of our adjunct faculty teaching the year-long general chemistry series embedded both the IDI-Classroom modules and the IDI-Lab modules in her courses and implemented them at NSC for three years. We had the opportunity to work closely with her to model the pedagogical strategies involved, demonstrate the assessment of student learning in this alternative context, and discuss the value of transforming early STEM. This adjunct faculty member ultimately earned a tenure track position at a nearby college and will incorporate the IDI approach at her new institution for years to come.

Despite these efforts, one of the more challenging aspects of carrying out this work is establishing an institutional or departmental culture that values a reflective and evidence-based approach to teaching (*31*). It is not unusual to have one or two faculty working towards excellence in STEM education while the rest of the department remains content with the status quo. In our case, many faculty and some administrators have embraced the shift in culture, yet we do not have full participation with the IDI curriculum within our own chemistry department. Research indicates that using change strategies that are "emergent" as opposed to "prescribed" can have an impact on successful faculty development (*32*). One example is encouraging a reflective practice through curriculum development as opposed to passive presentation of new ideas. Developing a shared vision towards change within a community as opposed to enacting policy around prescribed changes is another. Other research suggests that extensive support from disciplinary experts who are also trained in pedagogy can be essential in helping faculty both begin and sustain the use of alternative and more student-centered teaching strategies (*33*). In fact, without explicit training, even widely supported pedagogical approaches may not result in student learning gains (*34*).

Conclusion

The transformation of early STEM courses and programs is essential to creating meaningful learning experiences that have high impact and expand the outcomes we can achieve with our students at two year colleges. At NSC, we are currently addressing this by finding synergy between several independent projects. Combining efforts and finding overlap has provided a mechanism to develop a more holistic approach to transformation of our STEM program into a seamless package that is getting close to addressing all the essential elements of a successful STEM program (*1*). We have developed *innovative curriculum* in the form of interdisciplinary investigations for the classroom. We have created *early research experiences* that span first year CUREs, a second-year undergraduate research program, and an REU that was developed specifically for community college students across the region. Finally, we have launched an *ecosystem of extracurricular supports* that is focused on building meaningful relationships and providing the necessary resources for success.

Our work in these three areas has presented the opportunity to establish collaborations with other institutions, allowed us to obtain funding from a variety of sources both internal and external, and has presented numerous opportunities for professional development for STEM faculty. We continue to be challenged by the slow rate of change in academia and with institutional norms that can often inhibit progress. In the near future, we aim to continue to document the impact this work has had on student learning and student success. We are particularly interested in gaining a better understanding of how students develop into complex thinkers and seek to find more ways to capture student thinking as they work in both an interdisciplinary and a research context. In addition, we also aim to gain a deeper understanding of how early research and extracurricular supports impact the development of STEM identity in minoritized groups.

Acknowledgments

Interdisciplinary Investigations and CURE: This material is based upon work supported by the National Science Foundation under Grant No. 1432018. The authors are grateful for the tireless efforts of their co-PI's Dr. Heather Price and Dr. Anne Johansen, as well as the enthusiasm and support of their students and lab staff.

RST: This material is based upon work supported by the National Science Foundation under Grant No. 1643580, 0969607, and 0966314. The authors gratefully acknowledge the support of all the numerous investigators associated with these awards including Dr. Joshua Whorley, Jacob Ashcraft, Krystle Balhan, Dr. Thai Nguyen, Dr. Wendy Rockhill, Rick Downs, Mike Steffancin, Chad Hickox, and Peter Lortz.

REU: This material is based upon work supported by the National Science Foundation under Grant No. 1358835. The authors gratefully acknowledge the support and contributions of co-PI Marina Halverson and collaborator, Dr. Fritz Stahr.

References

1. President's Council of Advisors on Science and Technology. *Report to the President, Engage to Excel: Producing One Million Additional College Graduates With Degrees in Science, Technology, Engineering and Mathematics*, 2012.
2. Chen, X. *Students Who Study Science, Technology, Engineering, and Mathematics (STEM) in Postsecondary Education*; National Center for Education Statistics, Institute of Education Sciences, U.S. Department of Education, Washington, DC, 2009.
3. Chen, X. *STEM Attrition: College Students' Paths Into and Out of STEM Fields (NCES 2014-001)*; National Center for Education Statistics, Institute of Education Sciences, U.S. Department of Education, Washington, DC, 2013.
4. American Association of Colleges & Universities. *Achieving Systemic Change: A Sourcebook for Advancing and Funding Undergraduate STEM Education. The Coalition for Reform of Undergraduate STEM Education*, 2014.

5. American Association for the Advancement of Science. *Vision and Change in Undergraduate Biology Education: A Call to Action*, Washington, DC, 2011.
6. National Academies of Sciences, Engineering, and Medicine. *Integrating Discovery-Based Research into the Undergraduate Curriculum: Report of a Convocation*; National Academies Press, Washington, DC, 2015.
7. American Association of Community Colleges. *Fast Facts*, 2016. http://www.aacc.nche.edu/AboutCC/Pages/fastfactsfactsheet.aspx.
8. Freeman, S.; Eddy, S. L.; McDonough, M.; Smith, M. K.; Okoroafor, N.; Jordt, H.; Wenderoth, M. Active Learning Increases Student Performance in Science, Engineering and Mathematics. *Proc. Natl. Acad. Sci.* **2014**, *111*, 8410–8415.
9. Barr, D. A.; Gonzalez, M. E.; Wanat, S. F. The Leaky Pipeline: Factors Associated With Early Decline in Interest in Premedical Studies Among Underrepresented Minority Undergraduate Students. *Academic Medicine* **2008**, *83*, 503–511.
10. Hurtado, S.; Han, J. C.; Saenz, V. B.; Espinosa, L. L.; Cabrera, N. L.; Cerna, O. S. Predicting Transition and Adjustment to College: Biomedical and Behavioral Science Aspirants' and Minority Students' First Year of College. *Res. Higher Educ.* **2007**, *48*, 841–887.
11. Ong, M. Broadening Participation: The Status of Women of Color in Computer Science. *Communications ACM* **2011**, *54*, 32–34.
12. Mansilla, V. B.; Duraisingh, E. D. Targeted Assessment of Students' Interdisciplinary Work: An Empirically Grounded Framework Proposed. *J. Higher Ed.* **2007**, *78*, 215–237.
13. Loertscher, J.; Green, D.; Lewis, J. E.; Lin, S.; Minderhout, V. Identification of Threshold Concepts for Biochemistry. *CBE-Life Sci. Educ.* **2014**, *13*, 516–528.
14. Cooper, M. M.; Williams, L. C.; Underwood, S. M. Student Understanding of Intermolecular Forces: A Multimodal Study. *J. Chem. Educ.* **2015**, *92*, 1288–1298.
15. Owens, K. S.; Murkowski, A. J.; Price, H.; Johansen, A. *Using Interdisciplinary Investigations to Advance Student Understanding of Intermolecular Forces*, unpublished.
16. Mansilla, V. B. Interdisciplinary Learning: A Cognitive-Epistemological Foundation. In *Oxford Handbook for Interdisciplinary*, Oxford, U.K., 2016.
17. Murray, D. H.; Obare, S. O.; Hageman, J. H. Early Research: A Strategy for Inclusion and Student Success. In *The Power and Promise of Early Research*;Murray, D. H., Obare, S. O., Hageman, J. H., Eds.; ACS Symposium Series 1231; American Chemical Society: Washington, DC, 2016; pp 1–32.
18. Community College Undergraduate Research Initiative. *Investing in Impact*, 2014. http://www.ccuri.org/publications.
19. Kuh, G. D. *High-Impact Educational Practices: What They Are, Who Has Access to Them, and Why They Matter*; Association of American Colleges and Universities, Washington, DC, 2008.

20. Lopatto, D. *Science in Solution: The Impact of Undergraduate Research on Student Learning*; Council on Undergraduate Research and Research Corporation for Science Advancement, Washington, DC and Tuscon, AZ, 2009.
21. Kardash, C. M. Evaluation of an Undergraduate Research Experience: Perceptions of Undergraduate Interns and Their Faculty Mentors. *J. Educ. Psych.* **2000**, *92*, 191–201.
22. Lopatto, D. The Essential Features of Undergraduate Research. *Council on Undergraduate Research Quarterly* **2003**, *24*, 139–142.
23. Nagda, B. A.; Gregerman, S. R.; Jonides, J.; Von Hippel, W.; Lerner, J. S. Undergraduate Student-Faculty Partnerships affect Student Retention. *Review of Higher Education* **1998**, *22*, 55–72.
24. Brownell, S. E.; Kloser, M. J. Toward a Conceptual Framework for Measuring the Effectiveness of Course-based Undergraduate Research Experiences in Undergraduate Biology. *Studies in Higher Education* **2015**, *40*, 525–544.
25. Lopatto, D.; Hauser, C.; Jones, C. J.; Paetkau, D.; Chandrasekaran, V.; Dunbar, D.; MacKinnon, C.; Stamm, J.; Alvarez, C.; and Barnard, D. A Central Support System can Facilitate Implementation and Sustainability of a Classroom-based Undergraduate Research Experience (CURE) in Genomics. *CBE-Life Sci. Educ.* **2014**, *13*, 711–723.
26. Corwin, L.; Graham, J.; Dolan, E. Modeling Course-Based Undergraduate Research Experiences: An Agenda for Future Research and Evaluation. *CBE-Life Sci. Educ.* **2015**, *14* (1), 1–13.
27. National Academies of Sciences, Engineering, and Medicine. *Undergraduate Research Experiences for STEM Students: Successes, Challenges, and Opportunities*; The National Academies Press, Washington, DC, 2017. https://doi.org/10.17226/24622.
28. Owens, K. S.; Murkowski, A. J. A Model of Interdisciplinary Undergraduate Research Experiences at a Community College. In *Undergraduate Research at Community Colleges*; Cejda, B. D., Hensel, N., Eds.; Council on Undergraduate Research; Washington, DC, 2009; pp 19–31.
29. Owens, K. S.; Murkowski, A. J.; Price, H.; Johansen, A. A High Throughput Model for Course-Based Research Experiences for First Year of Chemistry and Biology. In *Course-based Research: Providing Opportunities for all Students to Learn Through Undergraduate Research*; Stylus Publishing: Laguna Woods, CA, 2018.
30. Wilson, Z. S.; Holmes, L.; deGravelles, K.; Sylvain, M. R.; Batiste, L.; Johnson, M.; McGuire, S. Y.; Pang, S. S.; Warner, I. M. Hierarchical Mentoring: A Transformative Strategy for Improving Diversity and Retention in Undergraduate STEM Disciplines. *J. Sci. Educ. Technol.* **2012**, *21*, 148–156.
31. Carlone, H. B.; Johnson, A. Understanding the Science Experiences of Successful Women of Color: Science Identity as an Analytic Lens. *J. Res. Sci. Teach.* **2007**, *44*, 1187–1218.
32. Fairweather, J. *Linking Evidence and Promising Practices in Science, Technology, Engineering and Mathematics (STEM) Undergraduate*

Education; Paper for the National Academies National Research Council board of Science Education, 2009.

33. Henderson, C.; Beach, A.; Finkelstein, N. Facilitating Change in Undergraduate STEM Instructional Practices: An Analytic Review of the Literature. *J. Res. Sci. Teach.* **2011**, *48*, 952–984.
34. Wieman, C.; Deslaurier, L.; Gilley, B. Use of Research-Based Instructional Strategies: How to Avoid Faculty Quitting. *Phys. Rev. Spec. Top. Phys. Educ. Re.s* **2013**, *9*, 023102.

Chapter 3

Improving Student Outcomes with Supplemental Instruction

Vicki Flaris*

Chemistry Department, Bronx Community College, 2155 University Ave., Bronx, New York 10453, United States
***E-mail: vicki.flaris@bcc.cuny.edu.**

Supplemental Instruction (SI) or Peer-Led Instruction is a nationally recognized model of effective academic support for students in high-risk, difficult courses. SI was implemented at Bronx Community College in select high-risk, credit bearing courses, defined as courses in which more than 25% of students earn final grades of D, F and W or U. These included Chemistry, Math, Biology and Nursing (Pharmacology) courses and courses in non-STEM disciplines of Accounting, Art, Communications, Criminal Justice, History, and Psychology. This chapter will focus particularly on the SI program in STEM courses from the Spring 2014 to Spring 2017 during which time student grades in General Chemistry improved by 28% demonstrating an effective strategy to improve student success.

Importance of SI for Increasing STEM Workforce Diversity

Supplemental Instruction (SI) is a nationally recognized model of effective academic support program for students in high-risk, difficult courses. As developed at the University of Missouri at Kansas City (UMKC) (*1*), SI is a series of weekly review sessions for students enrolled in difficult courses. SI is provided for all students who want to improve their grade, and student attendance at SI sessions is voluntary. Students gather with classmates to compare notes, discuss important concepts and develop strategies for studying the subject. Each session is guided by an SI leader, a student who has previously been successful in the course. When students who participated in SI are compared with those who do not, the National Center for Supplemental Instruction at UMKC found that students in SI supported classes earned higher mean final course grade averages, succeeded at higher rates, had lower attrition rates and higher graduation rates (*2*).

The SI model has sometimes been used synonymously with the Peer Led Team Learning (PLTL) model, but there are differences between these student support models. The PLTL model originated at City College of New York in the early 1990s to address the low success rate of students in general chemistry (*3*). Peer-led workshops were incorporated into the teaching of general chemistry by reducing the lecture instruction time from four hours to three hours. Peer mentors in the PLTL model are similar to SI leaders in the fact that they are identified students who have done well in the course previously. PLTL mentors meet with students immediately after the scheduled lecture class time and work with students to discuss, debate, and engage in problem-solving related to the course material. Conversely, SI sessions are not linked to classroom lecture and may be offered at various times to suit student scheduling needs. Results at City College of New York and other collaborating institutions using PLTL indicated improved student attitudes and performance in general chemistry and other courses.

Bronx Community College is a Hispanic-Serving Institution (HSI) comprising of 98% minorities and 57% women and has an opportunity to play an important role in addressing the STEM workforce diversity gap. This workforce gap is exacerbated by racial and gender barriers to participation in STEM careers. Hispanics, Blacks, and American Indians or Alaska Natives, who together comprise 27% of the U.S. adult population, represent only 11% of the science and engineering workforce and just 15% of science and engineering's highest degree holders (*4*). In the next five years, major American companies estimate they will need to add nearly 1.6 million STEM-skilled employees (*5*). BCC is working to provide students with the appropriate workforce skills to compete for these STEM-related careers.

The decision to implement SI aligned with the BCC Performance Management Plan, which stresses the importance of preparing students for success in remedial, credit bearing and gateway courses. Our goals are to increase student success rates in reading, writing and mathematics on exit from remediation, improve academic performance, increase retention, and facilitate students' timely graduation.

Implementation of SI at BCC

In the Spring of 2014, courses were transitioned to incorporate the SI model developed at the UMKC (*1*). BCC piloted SI in select high-risk, credit bearing courses. "High-risk" courses at BCC are defined as those in which more than 25% of students earn final grades of D, F and W or U. In addition to Chemistry, SI was piloted in other STEM courses in Math, Biology, and Nursing (Pharmacology) disciplines and in non-STEM disciplines in certain Accounting, Art, Communications, Criminal Justice, History and Psychology courses. In the first semester, SI was funded through multiple sources (grant funding, Office of Academic Affairs and department specific budgets) and is currently funded through Coordinated Undergraduate Education (CUE) funding from CUNY. The CUE office's goal is to enhance the undergraduate experience in support of student services. SI helps students to master the content of the course and develop and apply effective learning and study strategies. Not only will it enable students to perform better in selected courses, but the skills that they learn as a part of participating in SI sessions will be generalizable to other courses.

SI Leader Recruitment and Requirements

SI Leaders are recommended by the department chairs or instructional faculty and are current BCC students who received a final grade of "A/A-" in the course, have great interpersonal skills and show an ability to assist others, have been interviewed by the SI supervisor and have completed a two-day training workshop on campus. It is important for SI Leaders to have successfully completed the particular course with a strong understanding of the course content so they are able to help guide students through the semester.

The qualifications of an SI Leader include the following criteria:

Required criteria:

- overall 3.0 GPA or above (on a 4.0 scale)
- course content-competency (determined by course professor)
- good interpersonal and communication skills (determined by SI supervisor)

Preferred criteria:

- grade of 3.4 or above in the selected course
- prior enrollment with the professor who is to teach the selected course

Once a course begins, SI Leaders attend all lecture classes, take notes, read all assigned material and conduct two 75-minute SI sessions each week throughout the term where they use strategies learned through the training workshops and integrate how-to-learn techniques with what-to-learn content. In the SI sessions, SI Leaders guide activities such as review sessions, share notes, discuss readings and develop and use learning and studying strategies. The study sessions are held in a dedicated space called the "Learning Commons" and not in individual subject tutoring areas or in the classroom as with the PLTL model.

The SI Leader also meets regularly with the SI supervisor for debriefing sessions to discuss observations of the SI sessions, the creation and use of SI session handouts, the planning of SI sessions, use of a wide variety of learning strategies and to notify their supervisor about problems or potential problems. SI Leaders collect and track the students' progress and submit data reports twice a semester (midterm and end of semester) on session attendance and all course assessments (papers, quizzes, midterm exam, final exam and final course grade).

SI Leader Training

SI Leaders participate in a two-day intensive training workshop prior to the beginning of the semester where they learn about SI principles and techniques to support student learning. Figure 1 shows the elements of SI training.

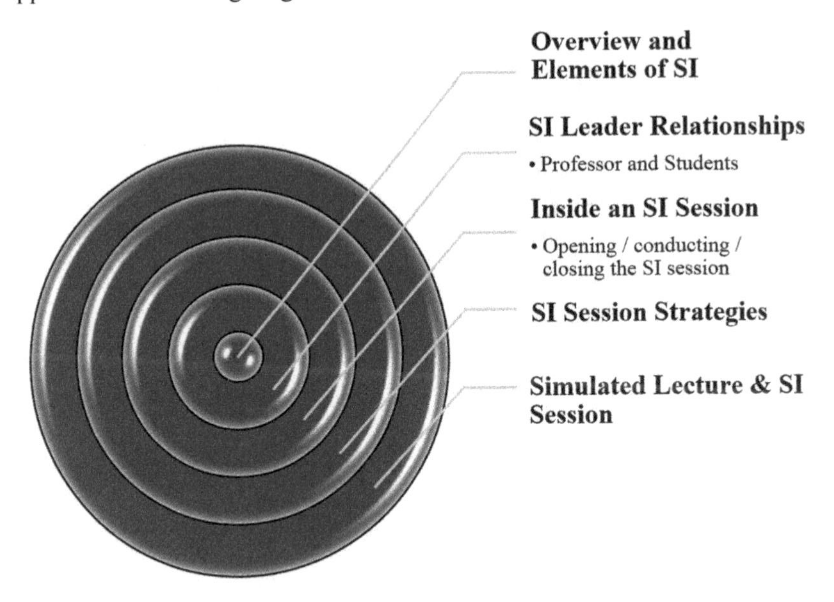

Figure 1. Components of SI leader training (6).

In addition to the two-day training workshop, SI Leaders also attend training sessions held throughout the semester. The SI Leader training includes a focus on helping students to develop effective problem-solving strategies for courses like chemistry, physics and mathematics which can be major obstacles for many students. Some of these learning strategies include:

- *Boardwalk Model*: This strategy is to assist students to build problem-solving skills. The SI Leader organizes the board into four columns and ask a number of volunteers to act as scribes for each column. The four types of information to be collected for each problem include; (1)

prerequisite knowledge; (2) mathematical steps; (3) narrative of steps; and (4) identification, solution or construction of a similar problem. This helps students who are solving problems such as percentage yield, Gas Law problems, concentration calculations etc. SI Leaders use this model when: (a) students don't know how to solve a problem; (b) students are stuck within a problem/solution; (c) to check students' understanding of how to solve each type of problem; or (d) to help organize and aggregate different types of problems.

- *Time Lines*: This is a technique which utilizes visual representation to improve processing the material. It is effective when trying to show a continuum of ideas and can be also used as a frame to which a student can add additional information. Timelines do not necessarily concern dates but may include processes. For example, the timeline for mitosis can show prophase/telophase at 0 minutes, prometaphase at 20 minutes, metaphase at 40 minutes, anaphase at 60 minutes and telophase at 80 minutes. Cytokinesis can be shown as a subset of Mitosis and a visual of the cell cycle may be shown below that.
- *First Line Only*: This strategy is used for students who need to be encouraged to take the first step toward finding a solution. A variety of types of problems are presented so the student builds confidence in addressing the first level of problem-solving. Secondly, a strict time limit to solving these problems is enforced. For example, how would you solve a molarity question versus a molality question? The problems need to be solved within 20 minutes as this is the time frame provided in final exams.
- *Send a Problem*: This strategy involves students working together in pairs. One student will start a problem and then pass it to a neighboring student to complete the next step, and then they alternate. Each student has a minute to complete their step until all steps have been completed. This strategy is typically used in mathematical based problems, such as calculus, solving limit type questions.
- *Vocabulary Activities*: These strategies are important in courses that use discipline-specific language. An SI Leader can create a vocabulary matrix and have students work together to fill in the matrix. Important terms will be listed from a chapter in the first column, the meaning of the term is placed in the second column, an example from the lecture notes is given in the third column, an example from the text is placed in the fourth column and the students place a new example in the last column. In the sciences students learn to use words like: *reliability, verifiability, clarity, empirical evidence, correspondence with natural laws, research methods and graphic presentation.* Science instructors tend to simplify complex ideas, while humanities instructors tend to favor probing complexity. SI Leaders for humanities, in contrast, encounter words like: *ambiguity, uncertainty, intuition, insight, self-knowledge, truths, process, symbolic representation.*
- *Big Idea*: Each student is asked to describe what he or she thought was the most important concept or new information learned in their

particular lecture. This exercise assists student to identify and organize information in a way that they are not overwhelmed by the sheer volume of information.

- *One Minute Paper*: The student is asked to write for one minute on a topic the SI Leader has covered. This activity helps students realize what they know and do not know. The students each hear each other's responses and the SI Leader then encourages students to discuss their similarities and differences.

In addition to the required training workshops, SI Leaders meet regularly with the SI supervisor and faculty instructor throughout the semester. The SI Leader prepares a written plan for each SI session which he/she then shows the supervisor and the faculty instructor in order to receive feedback. This plan includes learning objectives, difficult content and strategies for learning.

Benefits of Student Participation in SI Sessions

Overall it has been shown that regular attendance at SI sessions increases chances of earning better grades if a minimum of eight SI sessions have been attended, and students develop a better understanding of course content and effective ways of studying which are transferrable skills and useful in other courses (*6*). Since SI Leaders are often assigned to courses related to their majors, SI Leaders develop leadership skills and enhance their own academic skills and understanding within their discipline.

Student Success Data on Courses With and Without SI

In Table 1, data on student attendance verses non-attendance at SI sessions with final course grades for General Chemistry I (CHM 11) are shown (*6*). The means have been calculated by weighting the average of each semester with respect to the number of students in each sample.

Students attending SI sessions report them to be a helpful addition to the class according to post-survey results, which are not included here. The extra time each week in SI sessions allowed students to discuss material and their challenges with a peer SI Leader. Scheduling conflicts were the principal reason reported for low attendance at SI sessions. Since SI sessions are currently arranged around the SI Leader student course schedule, this may not be at a time convenient to the supported class schedule.

Examining the aggregate outcomes of several sections from the initial implementation of SI in Spring 2014 until Spring 2017, the conclusion was that in the General Chemistry I course (CHM 11), there was a 28% increase in mean final course grade of the students attending SI sessions versus those not attending. Although the student participation data is not shown here, similar increases in student final course grades were also observed in other STEM classes. In Mathematics for Probability and Statistics (MTH 23), there was a 17% increase, for Pre-Calculus Mathematics (MTH 30) there was a 16% increase in the mean final course grade, and in Pharmacology Computations (PHM 10) there was a 27% increase in the mean final course grade.

Interestingly, courses in non-STEM disciplines that implemented SI initially showed larger gains in student success compared to STEM courses. This may be attributed to the differences in the number of student contact hours and content coverage in these courses. Non-STEM courses typically meet for 3 hours a week while STEM courses meet five to six hours a week with lecture, recitation and lab components. Having the same amount of SI session hours to support students in STEM and non-STEM courses may have a greater impact in the non-STEM courses. With this in mind, the length of SI sessions in STEM courses was increased from two 75-minute session to two 90-minute sessions.

Overall, there was a 28% increase in mean final course grade in the General College Chemistry I course (CHM 11) of the students attending SI session versus those not attending from Spring 2014 until Spring 2017 (Figure 2). In Probability and Statistics (MTH 23), there is now a 37% increase and for Pre-Calculus Mathematics (MTH 30) there is a 13% increase in the mean final course grade, in Pharmacology Computations (PHM 10) there was a 28% increase in the mean final course grade and in Human Anatomy and Physiology I (BIO 23) there is a 39% in mean final course grade.

When comparing our STEM versus our non-STEM majors, larger increases have been achieved recently in the STEM majors where the college has put focus on retention of these students. The increases in final course grades are now comparable for students taking STEM courses versus non-STEM courses.

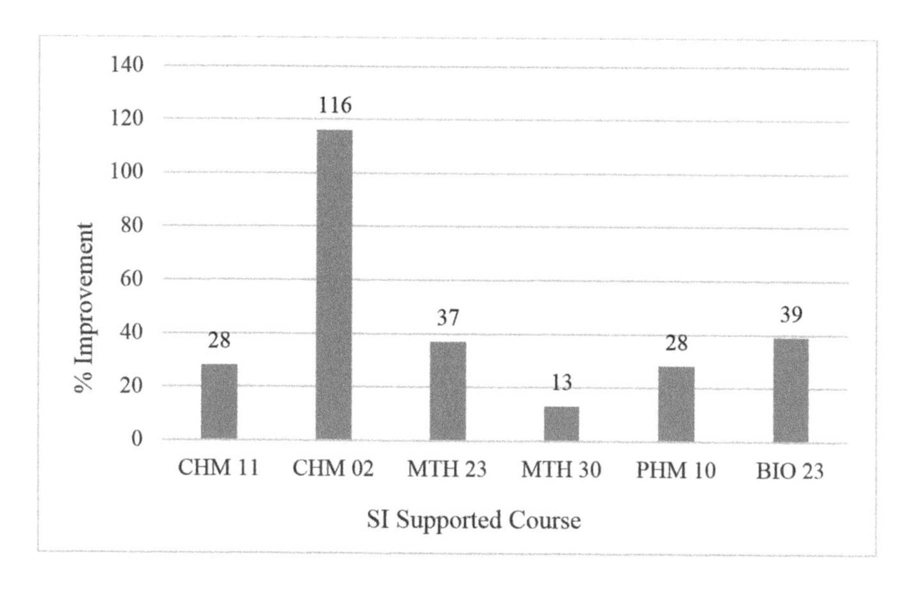

Figure 2. Percent improvement in final grades with SI for STEM courses: CHM 11 General Chemistry I; CHM 02 Remedial Chemistry; MTH 23 Probability & Statistics; MTH 30 Pre-Calculus Mathematics; PHM 10 Pharmacology Computations; BIO 23 Human Anatomy & Physiology I.

Table 1. SI Implementation in CHM 11 General Chemistry I

Semester	*Class size*	*Number of Attendees at SI Sessions*	*Mean # sessions students attended*	*Mean Final Grades*[*]		*Class Final Avg*
				SI Attendees	*Non-SI Attendees*	
Spring 2014	27	11	7	1.7	2.5	2.2
Spring 2015	22	3	7	2.7	2.4	2.5
Spring 2015	21	6	3	3.0	1.4	2.0
Fall 2016	64	38	5	2.6	1.6	2.0
Spring 2017	20	17	4	1.7	0	1.7
Spring 2017	24	11	3	2.7	1.5	2.1
Means				**2.3**	**1.8**	**2.1**

[*] All final grades out of 4.0.

Future Direction of SI

Student grades in STEM courses were shown to improve by significant amounts from the initiation of the SI program in Spring of 2014 through the Spring of 2017. There was a 28% improvement in Introductory Chemistry classes, a 37% improvement in Math and a 39% improvement in Biology. These increases in final course grades demonstrate that SI is an effective strategy to improve student success. Because of the student demographics at BCC, the SI model also contributes to the support of underrepresented students in STEM courses.

Moving forward, we are increasing our SI course offerings so there is a stronger focus on STEM courses. There will be more sections with SI Leaders provided and the SI Sessions for STEM courses will remain at 90 minutes per session. Initially we thought this increase in time would be appropriate as there was a larger gap in improvement in student grades in our non-STEM courses versus our STEM courses. Over recent years this gap has narrowed and this is also attributed to the quality of faculty teaching and to the more focused training of SI Leaders. The longer SI session time also helps to strengthen the problem-solving component skill sets of our students, a necessary component for STEM courses.

An increased focus will also be placed on marketing SI sessions in order to engage students in SI sessions earlier in the semester. There also will be more focus to coordinate SI sessions with the student's schedule to encourage student participation in an SI session immediately after their class to reduce knowledge decay (*8*, *9*). Wheeler conducted experiments investigating recall following two study conditions: one repeated test where a study trial was followed by multiple recall trials and a second repeated test where multiple study trials were conducted with no tests. When the retention interval was extended, forgetting was much more rapid in the study condition. The authors concluded that study and test trials have different effects upon memory, with study trials promoting memory acquisition, and test trials enhancing the retrieval process itself, which protects against subsequent forgetting (*9*).

There are plans for a stronger collaboration among faculty and SI Leaders to promote associative versus discrete learning experiences (*10*) and to create opportunities for multiple retrievals of the same information based on Anderson's theory of memory (*11*). Associative learning takes place through experience, and these experiences teach us what benefits us and what is harmful to us. Discrete learning experience is a method of teaching in simplified and structured steps. Instead of teaching an entire skill in one instance, the skill is broken-down and "built-up" using *discrete* trials that teach each step one at a time.

A future goal for STEM SI Leaders and faculty will be additional training to help students intentionally build their growth mindsets. Effective student and faculty practices will be identified to help support academic achievement and persistence of STEM students.

References

1. *Basic SI Model.* https://www.umkc.edu/asm/forms/si-article.pdf (accessed March 25, 2018).
2. *Supplemental Instruction Fall 2002 - Spring 2013 (SI) National Data.* https://www.umes.edu/uploadedFiles/_WEBSITES/CAAS/SI/Content/si%20national%20data%20updated%20slides_09-13-2013.pdf (accessed March 25, 2018).
3. Woodward, A.; Gosser, D. K.; Weiner, M. Problem Solving Workshops in General Chemistry. *J. Chem. Ed.* **1993**, *70*, 651.
4. *NSF Science & Engineering Indicators 2018.* https://www.nsf.gov/statistics/2018/nsb20181/; ac29 (accessed March 29, 2018).
5. *Business Roundtable / Change the Equation Survey on U.S. Workforce Skills* (December 3, 2014). https://www.ecs.org/wp-content/uploads/2014-BRT-CTEq-Skills-Survey-Slides_0.pdf (accessed March 29, 2018).
6. Flaris V., Bailey K. *Improving Student Outcomes with Supplemental Instruction.* Presented at the 251st National meeting of the American Chemical Society, San Diego, CA, March 2016.
7. The Curators of the University of Missouri. *The Leader's Guide to Supplemental Instruction*, 2015.
8. Bunce, D. M.; VandenPlas, J. R.; Soulis, C. Decay of Student Knowledge in Chemistry. *J. Chem. Educ.* **2011**, *88*, 1231–1237.
9. Wheeler, M. A.; Ewers, M.; Buonanno, J. F. Different Rates of Forgetting Following Study Versus Test Trials. *Memory* **2003**, *11*, 571–580.
10. Hockley, W. E. Item Versus Associative Information: Further Comparisons of Forgetting Rates. *J. Exp. Psychol.* **1992**, *18*, 1321–1330.
11. Anderson, J. R.; Bothell, D.; Byrne, M. D. An Integrated Theory of Mind. *Psychol. Rev.* **2004**, *111*, 1036–1060.

Chapter 4

Using Strategic Collaborations To Expand Instrumentation Access at Two-Year Colleges

Christopher J. Stromberg,[*,1] Debra Ellis,[2] Perry A. D. Wood,[2] Kevin H. Bennett,[1] Garth E. Patterson,[3] and Christopher Bradley[3]

[1]Department of Chemistry and Physics, Hood College, 401 Rosemont Ave., Frederick, Maryland 21701, United States
[2]Department of Science, Frederick Community College, 7932 Opossumtown Pike, Frederick Maryland 21702, United States
[3]Department of Science, Mount St. Mary's University, 16300 Old Emmitsburg Road, Emmitsburg, Maryland 21727, United States
[*]E-mail: stromberg@hood.edu.

This chapter describes a collaborative project between Hood College, Frederick Community College, and Mount St. Mary's University. The three institutions partnered together to acquire portable Fourier-transform, Raman, ultraviolet-visible, and X-ray fluorescence spectrometers that could be shared between them. This model of sharing instrumentation allowed students at all three schools hands-on access to a wider range of instrumentation than was previously available. Over 40 laboratory experiments have been developed between the three institutions, improving the educational experience for all students. Both the challenges faced and the lessons learned from developing and implementing the project are discussed.

Introduction

One of the biggest challenges of teaching chemistry at the post-secondary level is ensuring that laboratory experiences are teaching the students relevant skills (*1–8*). increasingly, this means providing access to modern instrumentation, even in courses such as general and organic chemistry (*2–4*, *6*, *9–12*). When students experience advanced instrumentation early and repeatedly, they learn these techniques at a deeper level, moving beyond seeing the instruments as "black boxes" to truly understanding what the instruments are telling them about their samples—and, by extension, the physical world around them (*1*, *2*, *5*, *6*, *8*, *9*, *13*). Access to modern instrumentation is important for promoting student interest in the sciences, as well as for preparing students for the workforce or for success after transferring to four-year schools (*1*, *4–6*, *8*, *9*, *12*, *14*). Such motivational factors may be as important for students as learning any specific scientific concept (*8*).

Unfortunately, providing access to instrumentation is expensive, and small, teaching-focused colleges struggle to purchase and maintain modern instruments (*2*, *4*, *6*, *11*, *15*, *16*). This problem is particularly acute at two-year colleges, where limited institutional budgets often mean that purchasing or maintaining instruments is a low priority. Some two-year colleges have been able to find funding for instrumentation to support teaching and research, but many others struggle to provide their students access to even basic instruments such as Fourier-transform infrared spectrometry (FT-IR) and nuclear magnetic resonance (NMR) spectrometers (*2*, *11*, *15–18*). In the case of NMR, advances in miniaturization have reduced the price of the instruments, but these costs are still out of reach for many small colleges (*19*).

Without access to these instruments, students at two-year colleges are at a distinct disadvantage. For students going into the job market directly after completing a two-year degree, instruction in instrumental methods are a fundamental part of their education, as instrumental methods are essential tools in virtually all industrial or research labs (*5*, *12*, *14*). For students transferring to a four-year college or university, upper-level courses often assume experience with instrumentation, so transfer students end up having to play catch-up relative to the students who began their studies at that institution (*16*, *20*).

Partnerships between institutions can be used to leverage and improve student access to modern instrumentation. This chapter describes one such partnership, between Frederick Community College (FCC), Mount St. Mary's University (MSMU), and Hood College. This partnership was funded through a grant from the National Science Foundation's Improving Undergraduate STEM Education (NSF IUSE) program, Award DUE-1431522. The partnership, involving a two-year college and two four-year schools, allowed all three institutions to significantly expand their access to modern instrumentation and to utilize those instruments throughout their curricula.

This collaboration also produced significant secondary benefits, including building closer relationships between the faculty of the three institutions and increasing their familiarity with each other's programs. This has improved both the formal and informal advising available for students considering transferring to the four-year schools, easing these transfer pathways.

This chapter begins with a brief description of the three institutions involved. A number of models for sharing instrumentation are discussed, followed by an account of our experiences with sharing portable instrumentation and the lessons learned that may be applicable to other partnerships.

Description of Institutions

The three institutions involved in this collaborative project are all located near Frederick, Maryland. Hood and FCC are both in Frederick itself (about 3 miles from each other). MSMU is located about 20 miles away, in Emmitsburg, Maryland.

FCC is a two-year institution that has served the Frederick area since 1957. FCC has 6,252 students attending for credit; 56% are female and 44% male. FCC offers 85 degree and certificate programs, with STEM being the fourth largest degree program. The Department of Science has 13 full-time faculty members in biology, chemistry, physics, and geology disciplines.

Hood College is a four-year, primarily undergraduate, liberal arts institution with 1,174 undergraduate students in 33 majors. Hood also has 30 graduate degree and certificate programs with 970 graduate students. Hood was founded as a women's college. In 1971, Hood allowed male commuter students to attend and went fully coeducational in 2003. Hood's undergraduate population is 62% female. The Department of Chemistry and Physics houses five full-time faculty members (four chemists and one physicist). The department offers majors in chemistry and biochemistry and minors in chemistry and physics.

Mount St. Mary's University is the second-oldest Catholic university in the United States, founded in 1808. MSMU is a four-year liberal arts institution with 1,653 undergraduate students in 40 major programs; 55% of their students are female. MSMU also has 383 graduate students in 18 master's degree and certificate programs. The Department of Science at MSMU offers majors in biochemistry, biology, chemistry, environmental science, and health sciences, along with minors in biology, chemistry, and environmental studies. The Division of Chemistry includes five tenure-track faculty members.

The three institutions have a history of working together and sharing resources—sometimes officially, sometimes on an *ad hoc* basis. For instance, when Hood was starting its nursing program, Hood and FCC arranged for Hood students to use some of FCC's nursing facilities. During the summer of 2012, while FCC's science building was under renovation, they had no viable space for teaching organic chemistry. Hood made available their organic chemistry laboratory and classroom space, so that the FCC organic chemistry courses were taught on Hood's campus.

A variety of unofficial arrangements to share instrumentation between Hood and MSMU have evolved over the years, especially for research purposes. These arrangements have generally involved Hood or MSMU faculty bringing research students to the other campus for a day or two during a summer session to use instrumentation unavailable or down for repair at their own school.

Models for Sharing Instrumentation

Without access to modern instrumentation, students must learn about techniques such as FT-IR and NMR spectroscopy from books, lectures, static spectra, and/or computer simulations of spectra. All of these provide students some background and understanding of the techniques, but none generate student interest the same way as using the instruments themselves (*1*, *5*, *6*, *9*, *16*, *20–30*). Working with spectra from a library or simulated spectra is useful for students, but these are generally "ideal" spectra, with none of the impurities or other issues that students must deal with when working with real-life spectra (*20*, *21*).

Access to instrumentation also substantially supplements other experiences in lab. For instance, determining the success and purity of organic synthesis reactions (especially on the micro scale) is difficult without access to IR or NMR. With real spectra of their products, students can think more specifically about what might have caused unsuccessful or impure syntheses and modify their techniques next time (*20*, *21*, *31*).

Students also get additional satisfaction from analyzing spectra of compounds they have personally synthesized. Analyzing their own compounds gives them a sense of ownership of their work and motivates them to put more time and effort into understanding both the spectra and the physical processes underlying the instrumental techniques (*1*, *9*, *20*, *30*).

Four main models for collaborative sharing of instrumentation exist: sending samples to another institution for analysis, sharing instrumentation, utilizing web-enabled interfaces to remotely control instrumentation, and sharing portable instrumentation. Each of these models address some or all of the issues noted above. Each offers benefits and potential drawbacks, some of which are discussed in the next section.

Sending Samples Out for Analysis

One of the simplest models for getting access to real spectra of students' samples is to send the samples to another institution for analysis (*20*, *32*). This requires gathering samples, packing them up, and shipping them to another institution. At the other institution, someone (sometimes a laboratory technician, sometimes an undergraduate or graduate student) will run the samples and return the results to the students.

This model can provide students with direct feedback on the results of their synthetic work, and, if only raw data is provided (such as a free induction decay for NMR), the students can get experience processing and analyzing the data (*20*, *21*). The students will not, however, get any hands-on experience with the instrument itself or the sample preparation and loading required for the instrument. There will also be a delay in receiving the data, although this can be as little as a week, depending on the arrangements between the institutions.

Generally, there is a cost associated with analyzing the samples. These costs vary widely, starting at a couple of dollars per sample, depending on the services provided and the particular agreement between the institutions (*20*, *32*). Even a nominal fee will add up over time, however, if an institution is working to provide all students in organic chemistry access to multiple NMR spectra over the course of a semester.

Physical Sharing of Benchtop Instrumentation

Another model for sharing instrumentation is for an institution (usually the four-year partner) to purchase an instrument and allow a local two-year school access to the instrument for research or teaching purposes.

This model can work for limited uses, especially for research, when the use of the instrument is going to be occasional and only for a day or two. As mentioned above, the collaborating institutions in this project have shared instrumentation in this way for research purposes in the past, allowing access to instruments that were unavailable at one of the institutions. This type of sharing generally works because it is short term and can be easily scheduled around other uses of the instruments.

In terms of shared instrumentation for teaching, however, this model has serious disadvantages. Even with institutions that are physically close to each other, such as FCC and Hood, the logistical challenges of transporting students to the other campus on a regular basis or for numerous sections are daunting, and this model would not be sustainable over the long term.

One case where such physical sharing of instrumentation might be successful would be a case where a two-year and a four-year school are co-located at the same site (*33*). Such sharing of space seems to be increasingly common, as a way to reduce costs through shared resources (*34*). However, these situations are the exception, and for most partnerships, physical sharing of benchtop instrumentation is simply not feasible.

Web-Enabled Interfaces

Another common example of sharing instrumentation is through the use of web-enabled interfaces. Most high-end instruments come with software and protocols that allow the instrument to be controlled via a web interface. This means that the instrument, in theory, can be run by users anywhere across the country or across the globe. This allows users without physical access to the instrument to get at least some of the experience of running the instrument.

There are many variants of this model (*2*, *6*, *10*, *16*, *31*), too numerous to list in full here. A quick search of the NSF's Major Research Instrumentation grants in recent years will show dozens of projects using web-enabled features to share their instrument with users at other sites. In most cases, samples are shipped to the institution housing the instrument, and remote instrument time is scheduled, allowing the remote user to run the instrument to analyze their own sample. The host institution must provide at least some support for sample preparation and loading. In many cases, the host institution provides personnel to monitor the instrument while it is being operated remotely. Accessories such as autosamplers can reduce the amount of in-person supervision required (*2, 16, 31*), but some users have raised concerns about the reliability of these autosamplers (*20*).

One of the issues with this model of instrument sharing is the level of support required of the host institution. Some institutions are able to provide sufficient support using undergraduate or graduate students to prepare samples, load them, and monitor the instrument, while others use a lab technician for these duties (*2, 10*). Whatever the level of support offered, the institution hosting the instrument will incur ongoing costs associated with providing this support, and that institution must be willing to absorb those costs or find a way of offsetting them.

Another issue with this model is the complexity of scheduling required. Since remote sampling must be scheduled around in-person users, careful planning is required (down to 15 minute intervals, in some cases) (*2, 31*).

A final issue with this model is that, while it provides the user experience with running the software to control the instrument, some of the sample preparation and physical loading of the sample must be done on location, so the students do not get experience with these steps in the process. For some types of instrumentation, such as attenuated total reflectance FT-IR or gas chromatography/mass spectroscopy (GC/MS), these steps are trivial, and the students will not lose much by not doing them. However, for other instruments, such as X-ray crystallography, sample preparation and mounting/centering of the crystal in the diffractometer are critical to acquiring good data, and experience with these steps is an important part of knowing the technique.

Sharing Portable Instrumentation

A final model (and the one used in this partnership) is to purchase portable or hand-held instrumentation that can be shared between the partner institutions (*4, 12, 35*). The big advantage of portable instrumentation is that the instrument itself can be moved between sites, allowing the instrument to be brought to the students, rather than trying to bring the students (physically or virtually) to the instrument. This means that students get direct, hands-on access to the instrument and, when they use the instrument, they get experience with all aspects of the instrument's operation. They also get their results in real time.

Planning the use of the instruments is still an issue when sharing portable instrumentation, but the scheduling between institutions is on a weekly or monthly basis, rather than fractions of an hour. This does require coordination of class and lab schedules before the beginning of each term, to make sure that each institution will have the instruments available when they are needed. Given the close collaboration in the partnership described here, this communication and scheduling has not been an issue, even when need for instruments has come up on short notice.

With the close physical proximity of the three institutions in this partnership, moving the instruments between institutions is very easy. Generally, a faculty member will simply pack up the instrument and drive it to the next institution, allowing the instruments to be relocated frequently, if necessary. This model of sharing portable instrumentation could work with larger distances between institutions, but the barrier of activation to shifting the instruments around would be higher.

Another potential issue with portable instruments is physical security. Because the instruments are, by definition, portable, they can be stolen more easily than benchtop instrumentation. Thus, a way to secure the instruments when they are being used—and secure, lockable storage for when they are not in use—is important at each site.

Portable instruments generally have lower resolution and sensitivity than their benchtop counterparts, but for teaching purposes, this is usually not a significant issue. While a benchtop IR spectrometer might have a resolution down to 0.125 cm^{-1}, for instance, the Bruker Alpha instrument used in this partnership has a resolution of 2 cm^{-1}. For any solid or solution-phase work, this resolution is more than sufficient. The only issues arise when gas-phase samples are used, but these samples are unusual for experiments during the first two years of study.

One final benefit of portable instrumentation is that they can be carried into the field. Portable instruments are ideal for archaeological, geological, or environmental field work, allowing for the analysis of samples on-site, rather than transporting samples back to a lab for analysis. This opens up a wide range of potential uses that would be difficult or impossible with benchtop instrumentation.

Development of the Partnership

While the aforementioned institutions have a history of working together, this particular partnership had its roots in the sabbatical project of one of the authors (Debra Ellis) during the spring of 2011. One part of Dr. Ellis' sabbatical work involved talking with local two-year and four-year colleges to find ideas for improving the organic chemistry sequence at FCC.

While visiting Hood, Dr. Ellis and Dr. Stromberg met to discuss challenges faced by their institutions. At the time, they identified three major issues that they had in common: a lack of resources, the difficulty in finding qualified adjunct instructors, and the struggle to provide students with modern instrumentation.

Of these challenges, the one that seemed most amenable to a collaborative solution was the last: providing instrumentation for our students. At that time, FCC had very limited instrumentation, including a set of outdated Spectronic 20 single wavelength visible spectrometers and three new melting point setups. Hood has had better success getting instrumentation, in large part thanks to funding through NSF's Instrument and Laboratory Improvement (ILI); Course, Curriculum, and Laboratory Improvement (CCLI); and Transforming Undergraduate Education in STEM (TUES) programs. However, keeping their instruments up-to-date and purchasing new instruments has been a significant challenge. These discussions paralleled those between Dr. Stromberg and Dr. Bradley at MSMU.

The team considered the different models for shared instrumentation discussed above. These considerations, paired with an already-existing interest in portable instrumentation, particularly for fieldwork in archaeology, led the group to focus on the model of sharing portable instrumentation.

The next step was to identify which types of instruments would be the most beneficial for the institutions. A portable FT-IR was quickly chosen. Both four-year institutions already had benchtop FT-IR instruments, but FCC did not have any FT-IR capabilities, and these were seen as critical to improving the lab experience at FCC, especially in organic chemistry.

The only UV-Vis analysis then available to FCC was using outdated Spectronic 20 instruments no longer supported by the manufacturer. The team decided that having a portable, full-spectrum ultraviolet-visible (UV-vis) spectrometer (as opposed to the single-wavelength Spectronic 20s) would allow for additional improvements in the FCC laboratory curriculum, so this type of instrument was included, as well.

Hood had experience with a benchtop Raman instrument through a previous NSF CCLI grant (DUE-0632829), but neither MSMU nor FCC had Raman capabilities. Given the complementary nature of IR and Raman spectra and the availability of good portable Raman instruments, a Raman spectrometer was added to the list.

Finally, portable X-ray fluorescence (XRF) instruments have become increasingly important in both environmental (*36–40*) and archeological (*36, 41–44*) field investigations. This elemental analysis technique was not available to any of the three institutions, and the team decided that it would add significantly to the capabilities of their respective programs.

NSF IUSE Grant Identified

The original plan was to submit a proposal to fund the collaboration and purchase of the instruments through the NSF's TUES program for the anticipated May 2013 deadline, but that program was discontinued after the 2012 application cycle. Thus, planning was put on hold while we sought out other potential sources of funding.

In November 2013, the NSF released a new program solicitation, introducing the Improving Undergraduate STEM Education (IUSE) program, with a due date of February 4, 2014. The goals of this program seemed to align with the goals of the collaboration under discussion, so we decided to put a proposal together, despite the abbreviated time frame. Dr. Stromberg was identified as the PI for the project, and Drs. Bennett (Hood), Ellis (FCC), Wood (FCC), and Bradley (MSMU) were identified as co-PIs.

What followed was three months of considerable effort to formalize the collaboration, the project plan, and the instruments that would be requested.

While Hood and MSMU both had substantial experience with federal research proposals, including from the NSF, neither had ever proposed a collaborative project with other institutions. FCC had little to no experience with federal research proposals. Therefore, all three institutions were breaking new ground with this project.

Initially, the administration at FCC seemed resistant to pursuing a research proposal through the NSF. The initial attitude seemed to be that, since community college faculty are there to teach, why would they pursue a research grant? This resistance quickly disappeared, however, as the benefits of the project were explained, and the FCC administration quickly became very supportive of the proposal.

The lack of familiarity with federal research proposals did add some additional hurdles in gaining formal approval from FCC for the project to move forward. Challenges included uncertainty about who should be administratively responsible for the project, a lack of a negotiated indirect cost agreement, and a requirement that such a proposal be approved by the entire President's Council before it could be submitted. Fortunately, the project team was able to surmount these obstacles and met the submission deadline.

Despite assurances from Dr. Stromberg that such projects are rarely funded on the first submission and the team would have a chance to refine and revise the project plan and would not actually have to implement it right away the project was formally funded on July 30, 2014.

Putting Ideas into Practice

The formal start date of the project was January 1, 2015, but as soon as the official notification of the award came through, the project team began putting the project structure into place.

Instruments Purchased

The first step in carrying out the proposed project was to identify and purchase the instruments. For each type of instrument, a number of models from various vendors were considered and compared. The final selection was made based on both the price and capabilities of each instrument. See Table 1 for a list of the instruments chosen and their specifications.

One of the main considerations was to find instruments that would be highly flexible, so they could be used in a number of different educational and research contexts. Consequently, most of the instruments were purchased with additional accessories beyond the base model. The prices shown in Table 1 include these additional accessories.

The FT-IR and UV-Vis instruments were purchased by and are primarily housed at FCC, which has the primary responsibility for scheduling their use. The XRF and Raman instruments were purchased by Hood, and Hood is responsible for scheduling their use. Routine maintenance is the responsibility of the institution that owns each instrument. Costs of use are paid for by the users.

Table 1. Portable instruments selected for the collaboration

Type	*Manufacturer*	*Model*	*Specifications*	*Cost*
FT-IR	Bruker	Alpha	2 cm^{-1} resolution 375-7500 cm^{-1} wavenumber range Transmission sampling module Diamond ATR module Front-face reflectance module	$31,000
UV-Vis	Ocean Optics	Red Tide 650 UV-Vis	2 nm resolution 200-880 nm wavelength range	$ 3,100
Raman	SciAps	Inspector 500	1030 nm, 300 mW laser 10 cm^{-1} resolution 150-2450 cm^{-1} wavenumber range Point and shoot adaptor Microscope adaptor with XYZ stage 8 mm liquid sample vial adaptor 1 cm square cuvette holder adaptor	$51,000
XRF	Bruker	Tracer III-SD	Rh target X-ray tube 10 mm^2 silicon drift detector 145 eV resolution Vision camera Vacuum pump	$44,000

Project Structure

Once the instruments were purchased and installed, and initial training was completed, the team had to develop laboratory experiments, demonstrations, and other activities using each instrument. The intention was to incorporate these instruments across the curriculum at each school, so that students got repeated experience with each instrument in a variety of classes and contexts. This repeated exposure would deepen the students' knowledge of both the techniques and the fundamental principles behind them (*1*, *2*, *8*, *13*, *20*). In order to facilitate this, the grant provided each institution with funds to hire one or two student researchers for each of the next three summers, as well as to pay summer stipends to the project team to oversee the development of these activities.

The first summer of the project (2015), each institution would have one of the main instruments (XRF, Raman, and FT-IR). Their team of students and faculty would develop activities for that instrument during the summer, to be tested in their classes the following school year (or, in the case of FCC, during summer classes as the activities were developed). They would also be responsible for putting together instructions and training materials for the instrument. During the first summer, FCC worked with the FT-IR, Hood worked with the XRF, and MSMU worked with the Raman instrument.

Over the next two summers, the instruments rotated, so that each institution had a summer dedicated to activity development on each of the main instruments. Subsequent summers would also be used to review and revise already-developed experiences, based on the feedback from the initial test runs. Instructions, training documents, and developed experiences were shared among the institutions to reduce the amount of redundant effort and to help in fine-tuning the materials.

At the beginning of each summer, the entire team met for three days of additional training. These training days rotated among the three institutions, so each institution hosted one training day. During each training day, the faculty would provide general safety training, as well as information on the theoretical background and operation of the instrument being used at the host institution. Any safety concerns particular to the instruments were covered; in particular, laser safety training for the Raman instrument and radiation safety training for the XRF were emphasized.

The team met every two weeks during the summers. These group meetings also rotated among the three institutions, and the students at the host institution presented their recent work in a semi-formal talk, which included a PowerPoint presentation. The other students were expected to give informal updates on what they had been working on. These group meetings were open to any students doing research in the three departments, as well as any interested faculty or administrators, so there were often guests present in addition to the grant team. The host institution provided snacks, as we found that food was an important part of building community.

In addition to the group meetings, the team arranged a variety of social events throughout the summer. These included trips to local amusement parks, BBQs at faculty members' homes, a day on the Potomac River, and more. These social events were also a key aspect of building a cohesive group and were invaluable for building bridges between people at the different institutions.

Project Successes

The most obvious success of this project has been the curricular changes that have resulted at the three institutions. Each institution has added these instruments into their curricula at all levels, so several hundred students per year are getting hands-on experience with instrumentation that was previously unavailable. This change has been most pronounced at FCC, where the entire organic chemistry laboratory curriculum has been revised, adding instrumentation to virtually every experiment.

All told, more than 40 experiments have been developed for a variety of courses, Table 2. These courses range from non-majors courses designed for first-year students to upper-level courses for science majors. Activities for disciplines other than chemistry have also been developed, expanding the reach of the program.

Table 2. Experiments developed through the collaboration

Course	*Number of Experiments*
Forensics	1
Geology	1
Physics	1
General Biology	2
Non-majors Chemistry	4
General Chemistry	6
Organic Chemistry	20
Inorganic Chemistry	2
Quantitative Analysis	3
Instrumental Analysis	2
Quantum Mechanics	2
Thermodynamics	1

Funding for summer salaries, both for students to do the curricular development and for faculty to oversee the work, was critical to the success of this project. Time and money for professional development activities can be scarce at two-year schools (*15*), so this project provided a welcome opportunity for the FCC faculty to be able to focus less on the day-to-day of teaching and to think more broadly about improving the curriculum as a whole.

With the high teaching loads for faculty at teaching-focused institutions, gaining access to instrumentation will not guarantee its use in the curriculum. Time and effort needs to be devoted to developing good activities. Without compensation for that development work, it can easily end up low on the priority list.

The opportunity to do curricular development has been an especially important opportunity for the FCC students. Research opportunities have huge benefits for students at two-year schools, but such research opportunities are, unfortunately, rare (*15*, *17*, *18*, *40*, *45–48*). All told, seven FCC students worked on developing experiments for this project, six as summer research interns and one as an honors project. The opportunity to be involved in developing experiments that were immediately used in courses they had taken had a real impact on them (*49*). In the words of one of the summer interns, "Being free to ultimately take our work as far as writing an instructor and student lab manual made me feel like I contributed valuable work to my undergrad institution."

Another success of this project has been the use of the instruments in outreach activities. The instruments have been used at local and regional STEM events at the elementary, middle, and high school levels (although we have generally had a better response from the older students when using instrumentation at such events).

The team has also made a concerted effort to disseminate the results of this project. This work has been presented at local, regional, and national conferences, both by faculty and by the student interns. Participants have given talks, roundtable discussions, poster presentations, and hands-on instrument demonstrations. Conferences have included community college-focused events such as the Association of Faculties for Advancement of Community College Teaching (AFACCT) and the Maryland Collegiate STEM Conference as well as national conferences such as the American Chemical Society National Meeting and the Biennial Conference on Chemical Education (BCCE).

Money was included in the grant to fund travel to these conferences, and this provided additional professional development opportunities for all of the faculty involved. The faculty from FCC have had limited opportunities to attend professional conferences, especially larger regional and national events. Attending these conferences gave the faculty opportunities not only to talk about their own work, but also to learn what is happening at other institutions across the country and to bring back ideas that could be implemented at their schools.

Attending and presenting at regional and national conferences also had a big impact on the students' educational and career tracks. Students at teaching-focused colleges (both two-year and four-year schools) often do not get a sense of the broad range of opportunities available in the field of chemistry. Taking students to present at conferences gives them a fantastic opportunity to see the myriad options out there for them to pursue. The students come back excited about the possibilities and more motivated to pursue them. To quote one of the FCC summer interns who presented at an ACS National Meeting, "Presenting this work was the gateway to my future internship, and it shaped my educational path."

As with many collaborative projects (*20*, *50*), one of the largest positive impacts of this collaboration has simply been the strengthening of the relationships between the three institutions. While the institutions had worked together in the past, contact between the faculty members of the institutions was relatively rare and on an as-needed basis. Since the beginning of this project, we have gotten to know each other—and the other institutions—much better.

Another major benefit has been that the faculty at FCC are now more familiar with the programs and transfer policies of the four-year schools, and they are better able to advise students interested in transferring to the four-year schools. This, in turn, benefits the students, as they get earlier and more accurate information about transferring, which can lead to fewer barriers and delays in the transfer processes (*51–55*).

Conversely, the faculty at the four-year schools have gained a better understanding of the challenges faced by potential transfer students, allowing them to advocate for policies at their institutions that could ease the transfer pathways for community college students. Making such policies more transfer-friendly can help eliminate barriers and allow students to transfer to and graduate from the four-year schools without unnecessary delays (*51–56*). One explicit example of this has been establishing degree-specific articulation agreements that have been developed for many of our programs.

This project has also laid a foundation for future collaborative efforts. For instance, Hood was recently awarded a grant from the NSF Robert Noyce Teacher Scholarship Program (Grant No. DUE-1660640). The Noyce program provides scholarships to students pursuing secondary education in STEM disciplines and includes an option to request additional funding for programs that include a community college partner. Hood's grant includes a collaboration with FCC to specifically recruit FCC students to participate in the program and receive the Noyce scholarships.

In the end, the collaboration has brought a number of direct benefits to all involved. It has resulted in improvements to the curricula; connectedness between two-year and four-year institutions, including articulation agreements allowing students to automatically transfer to the four-year institution; opportunities for students to present their findings in a supportive professional environment; and more.

It has also sparked tangential collaborations and developments. For example, the nature of this project was to utilize portable or transportable instrumentation to facilitate the physical sharing of hardware between institutions. Additional applications have been explored that could benefit from the use of portable instrumentation. As one example of this, MSMU has utilized a Griffin G450 transportable GC/MS to directly analyze the outgassing of volatile organic compounds (VOCs) from plant materials. This technique was used to differentiate between varieties of hops plant and could potentially be used to determine the health of the plant as well. The G450 incorporates a gas phase sampling interface that allows direct analysis of gaseous samples. This interface and the transportable nature of the instrument allowed for analyses that could not have been performed in any other way, since a hop plant may be as tall as 10 m and cannot easily be transported to a laboratory. Students have reported that experiments of this nature that clearly demonstrate why one might be interested in the analysis result in better learning outcomes and information retention. The same instrumentation has been utilized for *in loco* forensics analysis, also a popular topic with students.

Challenges along the Way

While this project has been extremely successful, it has not been without its challenges. None of the three institutions had previously participated in a collaborative project, so many of the procedures had to be developed along the way. This was especially true for FCC, which lacked experience with federal research grants. Consequently, the project team had to spend quite a bit of time developing procedures, such as those for time and effort reporting.

Since the four-year schools already had such procedures in place, the FCC staff were able to receive advice and guidance from their counterparts at the four-year schools. This sharing turned out to go both ways. For instance, FCC took Hood's time and effort reporting system as a starting point for their own system, and they adapted those procedures for FCC's context. In the end, they improved the procedures and forms so much that the Hood staff decided to revise their own system to incorporate the improvements from FCC.

In developing these procedures, we found that the administrative staff at our institutions were our best friends. Every time the grant team came up with a new issue or process for which there was no precedent, the professional staff went above and beyond to find ways to develop a process that worked. In particular, FCC's Assistant Director of Grants Management, Roseann Abdu, worked tirelessly to answer questions and develop needed processes. The involvement and buy-in from each school's professional staff were critical to the success of the project.

Another ongoing challenge was coordination between campuses. Each school has its own procedures for things like purchasing, reimbursements, and other administrative tasks. This occasionally led to some confusion and duplication of effort. Good communication between the faculty and staff at the different institutions has been essential. Initially, these differences in procedures tended to sneak up on us, but we quickly learned not to assume that everyone does things the same way and to ask about procedures before jumping in.

In addition to differing procedures, coordinating schedules among the three campuses has proved interesting. Each institution has its own academic calendar, its own timing for the beginning and end of semesters, and its own schedule for breaks. This made scheduling the summer work challenging to ensure the calendars worked for everyone. We also had to be flexible with the schedule to accommodate family vacations and other commitments for both the faculty and students.

Scheduling usage of the instruments has actually been easier than anticipated. As with other challenges, communicating early and often has been crucial, as well as maintaining flexibility when schedules changed. Unexpected opportunities to use the instrumentation have come up periodically. These have generally been able to be accommodated, even if they required additional shuttling of the instruments back and forth.

One of biggest obstacles for all of the faculty involved in the project was dealing with their heavy teaching loads. This was particularly an issue for the faculty at FCC, who generally teach during the summer sessions, which meant that they had to work around their teaching schedules to oversee their summer interns. In some ways, this turned out to be a benefit, because they were able to test the laboratory activities as they were being developed, since they had sections of the relevant courses running during the summers. On the other hand, a great deal of time and energy was required to fit everything in. The shortage of time was particularly acute during the fall and spring semesters, as none of the institutions provided any compensation or release time for project work during the school year.

Another issue that specifically affected FCC relates to the extensive usage of adjunct instructors. While the full-time faculty were interested and involved in the grant (even those not directly on the project team), the adjuncts had less buy-in and less familiarity with the equipment. Although training opportunities were developed to help the adjuncts feel comfortable with the instruments, some still did not use the instrumentation as extensively as the full-time faculty.

The high student turnover at FCC created additional challenges. At the four-year schools, one student intern generally returned from one summer to the next. The returning student interns were able to work significantly more independently and to help with training the new interns. At FCC, each set of student interns was new, requiring more training and oversight.

The turnover of students has also made it more difficult to get FCC students to present their work at conferences. At the four-year institutions, the summer interns were taking classes with us, so ensuring that they were meeting deadlines and working on their presentations was relatively easy. Most of the FCC interns have been second year students who were transferring to a four-year institution that fall. This made communicating with the students and monitoring their subsequent progress more challenging.

The final personnel-related challenge we faced was when one of the original co-PIs, Dr. Bradley, ended up accepting a new position and leaving MSMU in the final year of the grant. Dr. Bradley was able to find another MSMU faculty member, Dr. Patterson, to take over his role in the grant. The hand-off process was not entirely seamless, however. Better recordkeeping and sharing of documentation would have simplified the process and reduced the amount of reinventing the wheel that was necessary.

Conclusions: Lessons Learned

In carrying out this project, we have learned a number of lessons, as well as realized significant benefits. Some of these issues are mentioned above, but a few others deserve special mention.

If you are considering starting a collaborative project like this one, focus on building the relationships between schools and faculty organically BEFORE you start exploring the crazy ideas. The established relationships in this case made the hectic timeline for actually writing the proposal possible.

The discussions at the beginning of this project shaped the project into a true collaboration among equals. By exploring areas of common concern at the outset, the plan that evolved was tailored to meet the needs of all of the parties. This is only possible if everyone is involved in all stages of planning.

Grant agencies like the NSF often include inducements for working with partners from two-year colleges, including increases in the amount of money that can be requested. All too often, this leads four-year colleges to simply tack on a two-year partner to a project plan that they have already developed. Unfortunately, these tack-on relationships rarely benefit either institution as much as a true collaboration. Also, program officers and grant reviewers for agencies like the NSF can easily identify projects where the two-year school has simply been tacked on to get the extra money, and they do not look kindly on the practice.

The final take-home lesson from this project is this: the relationships built during the collaboration are likely to be the most important long-term outcome, so take the time to intentionally build those relationships. The group meetings over the summers were critical to making sure the project was kept on track, but the social events we had were probably more important to the long-term growth of the collaboration. Having unstructured, social time allowed us to talk about a wide range of issues. Out of those conversations came ideas not only for improving the current project but also ways to tackle other issues.

Acknowledgments

This material is based upon work supported by the National Science Foundation under Grant Nos. DUE-0632829, DUE-1431522, and DUE-1660640. Any opinions, findings, and conclusions or recommendations expressed in this material are those of the authors and do not necessarily reflect the views of the National Science Foundation.

In addition to the PI and co-PIs (the authors), the following faculty, staff, and students worked on projects related to the grant: ,

Frederick Community College
- Faculty and Staff
  - Dr. Wen Nellis
  - Patricia Sheppard
  - Edith Hillard
  - Emily Boward
- Students
  - Danielle Brown
  - Jerica Wilson
  - Melissa Gouker
  - Robin Carroll
  - Aaron Criner
  - Brianna Higgins
  - Thaina Brito
  - Pegah Avazpour
  - Laura Mundy

Hood College

Faculty

Dr. Susan Ensel

Students

Angela Mansfield

Alex Jarnot

Sarah Meyer

Erin Marshall

Mount St. Mary's University

Students

Annie Kayser

Meagan Suchewski

Michael Guckavan

References

1. Warner, D. L.; Brown, E. C.; Shadle, S. E. Laboratory Instrumentation: An Exploration of the Impact of Instrumentation on Student Learning. *J. Chem. Educ.* **2016**, *93*, 1223–1231.
2. Barot, B.; Kosinski, J.; Sinton, M.; Alonso, D.; Mutch, G. W.; Wong, P.; Warren, S. A Networked NMR Spectrometer: Configuring a Shared Instrument. *J. Chem. Educ.* **2005**, *82*, 1342.
3. Stewart, B.; Kirk, R.; LaBrecque, D.; Amar, F. G.; Bruce, M. R. M. Interchemnet: Integrating Instrumentation, Management, and Assessment in the General Chemistry Laboratory Course. *J. Chem. Educ.* **2006**, *83*, 494.
4. Smith, D. H. The Nebraska Instrument Sharing Consortium. *J. Chem. Educ.* **1986**, *63*, 68.
5. Miller, L. S.; Nakhleh, M. B.; Nash, J. J.; Meyer, J. A. Students' Attitudes toward and Conceptual Understanding of Chemical Instrumentation. *J. Chem. Educ.* **2004**, *81*, 1801.
6. Albon, S. P.; Cancilla, D. A.; Hubball, H. Using Remote Access to Scientific Instrumentation to Create Authentic Learning Activities in Pharmaceutical Analysis. *Am. J. Pharm. Educ.* **2006**, *70*, 121.
7. Bruck, L. B.; Towns, M.; Bretz, S. L. Faculty Perspectives of Undergraduate Chemistry Laboratory: Goals and Obstacles to Success. *J. Chem. Educ.* **2010**, *87*, 1416–1424.
8. Nakhleh, M. B.; Polles, J.; Malina, E., Learning Chemistry in a Laboratory Environment. In *Chemical Education: Towards Research-Based Practice*; Gilbert, J. K.; De Jong, O.; Justi, R.; Treagust, D. F.; Van Driel, J. H., Eds.; Springer Netherlands: Dordrecht, 2003; pp 69-94.
9. Richter-Egger, D. L.; Hagen, J. P.; Laquer, F. C.; Grandgenett, N. F.; Shuster, R. D. Improving Student Attitudes About Science by Integrating Research into the Introductory Chemistry Laboratory: Interdisciplinary Drinking Water Analysis. *J. Chem. Educ.* **2010**, *87*, 862–868.

10. Benefiel, C.; Newton, R.; Crouch, G. J.; Grant, K. Remote NMR Data Acquisition and Processing in the Organic Chemistry Curriculum. *J. Chem. Educ.* **2003**, *80*, 1494.
11. Webb, C.; Dahl, D.; Pesterfield, L.; Lovell, D.; Zhang, R.; Ballard, S.; Kellie, S. Modeling Collaboration and Partnership in a Program Integrating NMR across the Chemistry Curriculum at a University and a Community and Technical College. *J. Chem. Educ.* **2013**, *90*, 873–876.
12. Durick, M. A. The Study of Chemistry by Guided Inquiry Method Using Microcomputer-Based Laboratories. *J. Chem. Educ.* **2001**, *78*, 574.
13. McMinn, D. G.; Nakamaye, K. L.; Smieja, J. A. Enhancing Undergraduate Education: Curriculum Modification and Instrumentation. *J. Chem. Educ.* **1994**, *71*, 755.
14. Kenkel, J.; Rutledge, S.; Kelter, P. B. The Dupont Conference: Implications for the Chemical Technology Curriculum. *J. Chem. Educ.* **1998**, *75*, 531.
15. Brown, D. R. Undertaking Chemical Research at a Community College. *J. Chem. Educ.* **2006**, *83*, 970.
16. Cancilla, D. A. Initial Design and Development of an Integrated Laboratory Network: A New Approach for the Use of Instrumentation in the Undergraduate Curriculum. *J. Chem. Educ.* **2004**, *81*, 1809.
17. Gaglione, O. Underground Existence of Research in Chemistry in Two-Year College Programs. *J. Chem. Educ.* **2005**, *82*, 1613.
18. Petkewich, R. Community Colleges Tackle Research. *Chem. Eng. News* **2006**, *84*, 53–54.
19. Blümich, B.; Singh, K. Desktop NMR and Its Applications from Materials Science to Organic Chemistry. *Angew. Chem. Int. Ed.* in press.
20. Mills, N. S.; Shanklin, M. Access to NMR Spectroscopy for Two-Year College Students: The Nmr Site at Trinity University. *J. Chem. Educ.* **2011**, *88*, 835–839.
21. Vaughn, J. B. The Influence of Modern NMR Spectroscopy on Undergraduate Organic, Inorganic, and Physical Chemistry at Florida State University. *J. Chem. Educ.* **2002**, *79*, 306.
22. Chen, X.; Song, G.; Zhang, Y., Virtual and Remote Laboratory Development: A Review. In *Earth and Space 2010*; ASCE, 2010.
23. Hofstein, A.; Mamlok-Naaman, R. The Laboratory in Science Education: The State of the Art. *Chem. Educ. Res. Pract.* **2007**, *8*, 105–107.
24. Handelsman, J.; Ebert-May, D.; Beichner, R.; Bruns, P.; Chang, A.; DeHaan, R.; Gentile, J.; Lauffer, S.; Stewart, J.; Tilghman, S. M.; Wood, W. B. Scientific Teaching. *Science* **2004**, *304*, 521–522.
25. Michael, J. Where's the Evidence That Active Learning Works? *Adv. Physiol. Educ.* **2006**, *30*, 159–167.
26. Michael, J.; Modell, H. I., *Active Learning in Secondary and College Science Classrooms: A Working Model for Helping the Learner to Learn*; Taylor & Francis, 2003.
27. Romiszowski, A., The Development of Physical Skills: Instruction in the Psychomotor Domain. In *Instructional-Design Theories and Models: A New Paradigm of Instructional Theory*; Reigeluth, C. M., Ed.; Erlbaum: Mahwah, NJ, 1999; Vol. II, pp 457–481.

28. Tan, B. Teaching and Learning Physics in the 21st Century. *Innovation* **2012**, *11*, 8–12.
29. Wampold, B. E.; Wright, J. C.; Williams, P. H.; Millar, S. B.; Koscuik, S. A.; Penberthy, D. L. A Novel Strategy for Assessing the Effects of Curriculum Reform on Student Competence. *J. Chem. Educ.* **1998**, *75*, 986.
30. Malina, E. G.; Nakhleh, M. B. How Students Use Scientific Instruments to Create Understanding: CCD Spectrophotometers. *J. Chem. Educ.* **2003**, *80*, 691.
31. Kennepohl, D.; Baran, J.; Currie, R. Remote Instrumentation for the Teaching Laboratory. *J. Chem. Educ.* **2004**, *81*, 1814.
32. Murray, R. Shared Experimental Infrastructures. *Anal. Chem.* **2009**, *81*, 8655–8655.
33. Samuel, J.; Spackman, C.; Ruff, L.; Crucetti, J. J.; Chiappone, S.; Schadler, L. A Research University and Community College Collaboration Model to Promote Micro-Manufacturing Education: Preliminary Findings. *Procedia Manuf.* **2016**, *5*, 1168–1182.
34. LaSota, R. R.; Zumeta, W. What Matters in Increasing Community College Students' Upward Transfer to the Baccalaureate Degree: Findings from the Beginning Postsecondary Study 2003–2009. *Res. High Educ.* **2016**, *57*, 152–189.
35. Long, G. L.; Bailey, C. A.; Bunn, B. B.; Slebodnick, C.; Johnson, M. R.; Derozier, S.; Dana, S. M.; Grady, J. R. Chemistry Outreach Project to High Schools Using a Mobile Chemistry Laboratory, Chemkits, and Teacher Workshops. *J. Chem. Educ.* **2012**, *89*, 1249–1258.
36. Palmer, P. T. Energy-Dispersive X-Ray Fluorescence Spectrometry: A Long Overdue Addition to the Chemistry Curriculum. *J. Chem. Educ.* **2011**, *88*, 868–872.
37. West, M.; Ellis, A. T.; Potts, P. J.; Streli, C.; Vanhoof, C.; Wobrauschek, P. 2014 Atomic Spectrometry Update - a Review of Advances in X-Ray Fluorescence Spectrometry. *J. Anal. At. Spectrom.* **2014**, *29*, 1516–1563.
38. Butler, O. T.; Cairns, W. R. L.; Cook, J. M.; Davidson, C. M. 2014 Atomic Spectrometry Update - a Review of Advances in Environmental Analysis. *J. Anal. At. Spectrom.* **2015**, *30*, 21–63.
39. Bachofer, S. J. Sampling the Soils around a Residence Containing Lead-Based Paints: An X-Ray Fluorescence Experiment. *J. Chem. Educ.* **2008**, *85*, 980.
40. Turner, R. Using Technology to Create a Scientific Learning Community. *J. Chem. Educ.* **2001**, *78*, 717.
41. Sianoudis, I.; Drakaki, E.; Hein, A. Educational X-Ray Experiments and XRF Measurements with a Portable Setup Adapted for the Characterization of Cultural Heritage Objects. *Eur. J. Phys.* **2010**, *31*, 419.
42. Stuart, B., *Analytical Techniques in Materials Conservation*; John Wiley & Sons Ltd.: Chichester, UK, 2007; p 424.
43. Poupeau, G.; Le Bourdonnec, F.-X.; Carter, T.; Delerue, S.; Steven Shackley, M.; Barrat, J.-A.; Dubernet, S.; Moretto, P.; Calligaro, T.; Milić, M.; Kobayashi, K. The Use of SEM-EDS, PIXE and EDXRF for

Obsidian Provenance Studies in the near East: A Case Study from Neolithic Çatalhöyük (Central Anatolia). *J. Archaeol. Sci.* **2010**, *37*, 2705–2720.

44. Shackley, S. M., *X-Ray Fluorescence Spectrometry (XRF) in Geoarchaeology*; Springer: 2010; p 245.
45. Lau, L. K. Institutional Factors Affecting Student Retention. *Education* **2003**, *124*, 126–136.
46. Bauer, K. W.; Bennett, J. S. Alumni Perceptions Used to Assess Undergraduate Research Experience. *J. Higher Educ.* **2003**, *74*, 210–230.
47. Lopatto, D. Survey of Undergraduate Research Experiences (SURE): First Findings. *Cell Biol. Educ.* **2004**, *3*, 270–277.
48. Lopatto, D. Undergraduate Research Experiences Support Science Career Decisions and Active Learning. *CBE Life Sci. Educ.* **2007**, *6*, 297–306.
49. Bovill, C.; Bulley, C. J.; Morss, K. Engaging and Empowering First-Year Students through Curriculum Design: Perspectives from the Literature. *Teach. High. Educ.* **2011**, *16*, 197–209.
50. Ungar, H.; Brown, D. R. ChemEd Bridges: Building Bridges between Two-Year College Chemistry Faculty and the National Chemical Education Community. *J. Chem. Educ.* **2010**, *87*, 572–574.
51. Packard, B. W.-L.; Gagnon, J. L.; Senas, A. J. Navigating Community College Transfer in Science, Technical, Engineering, and Mathematics Fields. *Community Coll. J. Res. Pract.* **2012**, *36*, 670–683.
52. Townsend, B. K.; Wilson, K. "A Hand Hold for a Little Bit": Factors Facilitating the Success of Community College Transfer Students to a Large Research University. *J. Coll. Stud. Dev.* **2006**, *47*, 439–456.
53. Flaga, C. T. The Process of Transition for Community College Transfer Students. *Community Coll. J. Res. Pract.* **2006**, *30*, 3–19.
54. Carlan, P. E.; Byxbe, F. R. Community Colleges under the Microscope: An Analysis of Performance Predictors for Native and Transfer Students. *Community Coll. Rev.* **2000**, *28*, 27–42.
55. Hoffman, E.; Starobin, S. S.; Laanan, F. S.; Rivera, M. Role of Community Colleges in STEM Education: Thoughts on Implications for Policy, Practice, and Future Research. *J. Women Minor. Sci. Eng.* **2010**, *16*, 85–96.
56. Ungar, H.; Brown, D. R. Strengthening Relationships between Chemistry Faculties at Two-Year and Four-Year Institutions. *J. Chem. Educ.* **2010**, *87*, 885–886.

Curricular Innovations

Chapter 5

Development of a Pre-Professional Program at a Rural Community College

Shayna Burchett*[*] and Jack Lee Hayes

Chemistry Department, State Fair Community College,
3201 W. 16th Street, Sedalia, Missouri 65301, United States
[*]E-mail: sburchett@sfccmo.edu.

Many of the courses required for entrance to professional programs are offered by community colleges making community colleges a valuable asset to students seeking professional careers. State Fair Community College has recognized the opportunity to better serve and attract the pre-professional student population. Three of the strategies that have been implemented to improve the pre-professional transfer student experience are the design of advising pathways for the pre-professional students, the development of an Associate of Science (AS) in Chemistry, and chemistry course curriculum modifications.

Introduction

Medical doctors, pharmacists, physical therapists, optometrists, dentists, and veterinarians each require significant educational investment to enter their field. These students and their fields of study will be hereafter called pre-professionals and are a focused population with specific needs. The National Center for Educational Statistics reported that there were 97,000 master's degrees and 67,400 doctor's degrees conferred in the field of health professions and related programs in 2013-2014 (*1*). Although this number is less than 10% of the number of bachelor's degrees earned in the same time frame, it is a significant population worthy of dedicated attention. Though the professional specific courses must be completed at dedicated institutions during the final portion of the terminal degree, freshman and sophomore level prerequisite courses can be taken at nearly any academic venue. Community colleges can provide a successful avenue for tentative pre-professional learners to complete these foundational courses with the personal attention often found at a small institution.

This chapter will provide insight into the efforts of a rural community college to develop and sustain a pre-professional program. In particular, customized embedded advising, a new Associate of Scence degree in Chemistry and curriculum modifications will be discussed.

Advising Pre-Professional Students

A traditional Associates of Arts degree is typically too broad for most professional programs; students need to complete specific courses required for their program and receiving institution. A national in-house survey illustrates the diversity between pre-professional program course requirements (Table 1). Because of the variety between disciplines and often between receiving institutions within disciplines, advising pre-professional studenets is a challenge. Early identification of potential receiving institutions is critical to avoid taking courses that do not meet the requirements for a specific program.

Students need to complete a rigid course schedule including potentially more than 60 hours of specific required courses before application to a professional program in order to be competitive applicants. Like many small institutions, State Fair Community College (SFCC) does not have a student population that supports offering science major courses every semester. Juggling the task of matching a specific receiving institution's requirements, forecasting scheduling conflicts, and managing pre-requisite course sequencing for each individual pre-professional student is an arduous task that requires constant research and attention to detail.

Due to only having 77 full time faculty and nearly 5,000 students enrolled, SFCC invested in dedicated advisor positions. Sixteen individuals were hired to support student success through educational planning. To support these new advisors, SFCC Chemistry developed advising pathway templates for students intending to pursue careers in the health professions and related fields (Figure 1). While the credit load per semester is ambitious, in the authors' experience, applicants with 16-19 hour course loads and high GPAs tend to be the most successful applicants. Pathways were designed as starting points to develop individual educational plans that meet the needs of most receiving institutions with the understanding that the individual plan should be modified as needed to meet the student's goals. Special notes were made for known scheduling conflicts between math-biology courses and chemistry-physics courses which had to be taken in opposing years. SFCC Chemistry has transitioned all on ground courses to an eight week accelerated format which has allowed for students to complete two chemistry courses in a single semester; for example, General Chemistry I is completed during the first eight weeks and Organic Chemistry I is completed during the second eight weeks. By compressing the courses and allowing students to complete all four coursesover four eight-week consecutive sessions, the authors have found that students have better retention across the chemistry courses.

Table 1. Percent of Professional Programs That Require Selected Specific Courses across the United States

	Medical	*Pharmacy*	*Physical Therapy*	*Optometry*	*Dental*	*Veterinary*
Bio I	75.17%	87.14%	59.82%	95.24%	93.85%	93.33%
Bio II	72.48%	70.00%	41.96%	85.71%	92.31%	80.00%
Chem I	71.14%	98.59%	87.50%	100%	96.92%	93.33%
Chem II	68.46%	97.86%	87.05%	100%	96.92%	73.33%
Organic I	71.14%	98.57%	-	100%	96.92%	93.33%
Organic II	59.06%	98.57%	-	14.29%	87.69%	53.33%
Physics I	70.47%	77.14%	90.63%	95.24%	95.38%	93.33%
Physics II	68.46%	21.43%	90.63%	95.24%	93.85%	76.67%
Biochemistry	29.53%	28.57%	-	76.19%	66.15%	93.33%
English I	47.65%	-	19.20%	90.48%	87.69%	73.33%
English II	42.95%	-	-	85.71%	81.54%	60.00%
Calc I	5.37%	97.86%	-	90.48%	4.62%	10.00%
Calc II	2.68%	-	-	-	-	-
Statistics	14.77%	73.57%	-	85.71%	7.69%	40.00%
Cell Bio	2.01%	5.71%	-	-	-	3.33%
Genetics	2.36%	8.57%	-	-	-	33.33%

Continued on next page.

Table 1. (Continued). Percent of Professional Programs That Require Selected Specific Courses across the United States

	Medical	*Pharmacy*	*Physical Therapy*	*Optometry*	*Dental*	*Veterinary*
No Specific Prerequisites	22.82%	-	-	-	-	-
Number of Schools Surveyed	149	140	226	21	65	30

First Fall Term (with appropriate pre-requisites)			
Course No.	**Course Name**	**Credit Hrs**	**Notes**
CHEM 123	Gen Chem I w/lab	5	8 week course
CHEM 221	Organic Chem I w/lab	5	8 week course
ENGL 101	English Comp I	3	
MATH 120	Trig	3	Fall only, MATH 114 pre-requisite
SPTH 101	Public Speaking	3	
	Total Hours	**19**	
First Spring Term (must follow First Fall)			
Course No.	**Course Name**	**Credit Hrs**	**Notes**
CHEM 124	Gen Chem II w/lab	5	8 week course
CHEM 222	Organic Chem II w/lab	5	8 week course
ENG 102	English Comp II	3	
MATH 130	Calculus I	5	Spring only
	Total Hours	**18**	
Second Fall Term			
Course No.	**Course Name**	**Credit Hrs**	**Notes**
BIO 125	Bio I w/lab	5	Fall only
PHYS 105	Phys I w/lab	5	Fall only, can be PHYS 118
HLTH 101	Personal Health and Fitness	2	
GEN ED	FA, Hum, Lit, Soc Sci	3	Check transfer institution
GEN ED	American Institutions	3	
	Total Hours	**18**	MATH 131 encouraged
Second Spring Term			
Course No.	**Course Name**	**Credit Hrs**	**Notes**
BIO 124	Bio II w/lab	5	Spring only
PHYS 106	Phys II w/lab	3	Spring only, can be PHYS 119
PSY 101	Gen Psych	3	Check transfer institution
GEN ED	FA, Hum, Lit, Soc Sci	3	Check transfer institution
GEN ED	FA, Hum, Lit, Soc Sci	3	Check transfer institution
	Total Hours	**17**	
2 Summers and/or other terms			
Course No.	**Course Name**	**Credit Hrs**	**Notes**
BIO 121	Microbio w/lab	4	Check transfer institution
BIO 207	Anatomy w/lab	4	Check transfer institution
BIO 208	Physiology w/lab	4	Check transfer institution
BIO 210	Genetics w/lab	4	Check transfer institution
MATH 127	Statistics	3	Check transfer institution
GEN ED	FA, Hum, Lit, Soc Sci	3	Check transfer institution
	Total Hours	**22**	

Figure 1. Sample SFCC AA Course Sequencing Advising Pathway for Pre-Med/Opt/DDS students.

One of the challenges encountered is that some advisors have not experienced the coursework in programs for which they are advising. Though most of the advisors have at least an Associate of Arts degree, they are often responsible for multiple career paths (>10) without any personal experience in STEM pathways. This has presented a challenge to the dedicated advising program. Currently, this challenge is being mitigated by constant communications between advisors and volunteer STEM faculty.

Development of Associate of Science (AS) Degree in Chemistry

One of the metrics used to evaluate community colleges is completion rate. For pre-professional students, completing a traditional AA degree often leaves gaps in meeting their pre-professional prerequisites and burdens them with an excess of courses which transfer as electives to their professional program. Recognizing this challenge, SFCC developed an AS in Chemistry to better align transferring pre-professional student course sequencing. This degree was designed with a partner four-year institution to meet the most common requirements at the time for pre-professional students. AS in Chemistry degree course requirements are outlined in Table 2.

A challenge with the AS in Chemistry degree is the inflexible nature of the degree design. Due to the multipath nature of pre-professional programs, suggested/required prerequisites are not well aligned with the AS in Chemistry. For example, the AS in Chemistry degree requires Introduction to Biology with Lab while many pre-professional degrees require a General Biology I course with lab. Students also find the degree too restrictive in the current financial aid environment which does not support courses not required in a degree program. For this reason, SFCC is not experiencing as many student completeers as desired.

Chemistry Course Curriculum Modifications

The presence of medical programs departing from traditional course requirements and embracing competency based admittance supports the consideration of course curriculum modification. The paradigm shift to competency based admittance is reflected in the modification of SFCC Chemistry's curriculum to better match the Workshop Summary on Chemistry Undergraduate Education published by the National Academy Press (*2*). As a result, SFCC Chemistry fosters undergraduate research, emphasizes early introduction to organic and biochemistry, and includes medical topics in chemistry course content to provide relevance for the pre-professional students.

Undergraduate research in SFCC chemistry courses takes on multiple forms. Students in all chemistry courses are required to participate in an in-house science symposium during the final week of their course where they present their findings on either a literature-based research project or a laboratory investigation. Student poster projects on given topics are displayed in the department hallway throughout the semester. Part of the general chemistry course curriculum includes a student-driven investigation under the direction of faculty. During spring semesters, students are encouraged to present their findings at the Missouri Academy of Science Chemistry Symposium; this tradition began in 2007 and has continued every year with the most recent 2017 meeting with four groups of students giving oral presentations of their research (*3–6*).

Table 2. AS in Chemistry Degree Course Requirements

Course	*Credit Hours*
English Comp I	3
English Comp II	3
Public Speaking	3
Personal Health and Fitness	2
Introduction to Biology with Lab	5
*Calculus I**	5
*General Chemistry I**	5
*General Chemistry II**	5
*Organic Chemistry I**	5
*Organic Chemistry II**	5
*Physics I**	5
*Physics II**	3
American/National Government	3
Fine Arts, Humanities, Literature, or Social Sciences	9
Electives	3
Total	63

* Courses that match well with pre-professional program pre-requisites.

The first semester of general chemistry has been adjusted to include an emphasis on early introduction to organic and biochemistry topics. Subjects such as stoichiometry, equilibria, and spectroscopy are approached from an organic/biochemistry perspective through macroscale saponification, macroscale fermentation, steam distillation of essential oils, aspirin synthesis, UV-Vis and Raman product evaluation, and gas/column/paper/thin layer chromatography investigations. Several of these activities incorporate medical topics in the chemistry curriculum. In addition, molecular modeling of medications, dyes, polymers, proteins, and amino acids using both commercial molecule kits and found materials allow students to explore the structure-function relationship of molecules used in medicine.

Future Directions

SFCC is constantly seeking ways to improve the pre-professional student experience. To support the dedicated advising staff, chemistry faculty regularly monitor students who declare any pre-professional major and offer support through embedded advising in chemistry courses. In addition, faculty assist students in identifying potential receiving institutions and aid the development of customized advising pathways. As the customized advising pathways are established, they are shared with the advising staff as potential templates with the understanding that the pathways need to be frequently reevaluated to monitor for shifts in program requirements.

The AS in Chemistry degree is under review to identify flexibility in the existing degree program; flexibility is needed in the level of biology and physics course selections as the four semesters of chemistry are seldom the challenge due to most pre-professional programs requiring chemistry. Our institution is also investigating the potential to replace the degree with an AS in Pre-Professional Studies coordinated with a different four year partner institution.

Chemistry faculty regularly communicate with local professionals and evaluate courses each semester to identify opportunities to improve chemistry course curricula for pre-professional students. A current focus has been placed on expanding instrumentation providing students increased opportunities to experience analytical procedures and in turn enhancing undergraduate research.

Supporting a pre-professional program is a perpetual pursuit; professional programs undergo constant revision for improvement and so the pre-professional programs must follow suit. The authors intend to maintain a faithful vigil on professional programs and remain flexible in efforts to support the students who aspire to become our future health and medical professionals.

References

1. *Digest of Education Statistics*; NCES 2016-014; U.S. Department of Education, National Center for Education Statistics, 2015; Chapter 3.
2. National Research Council. *Undergraduate Chemistry Education: A Workshop Summary*; National Academies Press: Washington, DC, 2014.
3. Smith, A.; Warsawski, V. *Grow and Glow: Investigating Tissue Bioluminescence.* Presented at the Missouri Academy of Science, 2017.
4. Maher, J.; Stanford, C. *Bird Is the Word: An Analysis of Nutrients in Bird Seed.* Presented at the Missouri Academy of Science, 2017.
5. Smith, K.; Tempel, A. *Quantitative Water Analysis: A "Nu" Variant on Water Torture Using MicroLAB FASTSpec 528.* Presented at the Missouri Academy of Science, 2017.
6. Magras, A.; Miranda, J. *Raman Spectroscopy: The Verification of Acetylsalicylic Acid.* Presented at the Missouri Academy of Science, 2017.

Chapter 6

Student Affective State: Implications for Prerequisites and Instruction in Introductory Chemistry Classes

J. Ross,* C. Lai, and L. Nuñez

**Department of Chemistry, East Los Angeles College,
1301 Avenida Cesar Chavez,
Monterey Park, California 91754, United States
*E-mail: rossj2@elac.edu.**

The impact of students' affective state on success in entry-level chemistry classes is often underestimated or overlooked. Despite learning theory stating the importance of other variables such as students' affective state, instructors often hold true that students' cognitive and mathematical abilities are all that relate to mastering chemistry. The academic barrier to enroll in introductory chemistry classes is typically a passing grade in a mathematics prerequisite class. Many students will likely encounter affective domain barriers to content mastery in chemistry, which are not addressed by mathematics prerequisites. Driven by the long-term goal to improve student success and learning in entry-level chemistry classes, students' affective state was evaluated using a chemistry self-concept inventory. This work reveals action items that could positively impact students' achievement in chemistry.

Introduction

As instructors seek to improve student learning in chemistry, conversations centered on the cognitive aspects of learning chemistry typically surface. After all, chemistry is a natural science that is infused with a variety of mathematical techniques, abstract concepts, and academically obscure vocabulary. Charged with promoting higher levels of thinking in education in the 1950s, the cognitive, affective and psychomotor domains of learning were identified (*1*, *2*). The cognitive domain centers on an individual's mental skills while the affective domain relates to an individual's feelings and emotional state. The psychomotor domain concerns an individual's physical skills.

There is a significant link between a students' mathematical ability and achievement in chemistry (*3–7*). Therefore, it is logical that at the introductory college level, students typically must pass a prerequisite mathematics class prior to entering a chemistry class. Despite decades of educational research charting the composite structure of students' learning domains, chemistry education still often stresses only the cognitive aspect of learning and does not address the affective domain (*8*). Arguably, the psychomotor domain is catered by concomitant laboratory activities. Scholars of education and learning have known and disseminated the links between academic achievement in a discipline and students' attitude toward the objective subject of the discipline for some time (*9–17*). While controlling for the cognitive aspects of attitude, research has shown the positive role played by the affective component of students' attitude toward the conceptual development of science (*18*, *19*).

Despite the abundance of critical evidence supporting the role of students' affective domain in science learning, curriculum development efforts in science education remain reluctant to include effective assessments of students' attitude toward science (*8*). Many reasons for this exclusion of attitude assessment abound and center on a misinterpretation of what is meant by "attitude". Often, students' attitude is viewed as a unidimensional construct that can be accessed simply by administering a few survey questions. This is contrary to what learning theory has to say about students' attitude (*10*, *20–22*).

Rather than being an unidimensional construct, as would appear from a naïve inspection of Bloom's work (*1*, *2*), students' affective domain is a complex multitude of distinct yet related factors, which can be appropriately monitored when adequately identified and acknowledged. Among the unique sub-constructs of student affect are self-concept, self-efficacy, and attitude. Self-concept is an introspective component of the affective domain wherein a student reflects upon and evaluates his/her self with respect to past experiences with a body of knowledge (*12*, *17*, *19*, *23–27*). Self-efficacy is another introspective component of the affective domain wherein a student evaluates his/her prospects for performing a specific future task within a content area (*17*, *24*, *28–33*). Attitude is a learned predisposition to act favorably or unfavorably toward an object (*22*, *34–45*).

Studies have shown that students' chemistry self-concept is a construct that positively correlates with and can be used to predict chemistry achievement, although it is more than just a predictor of achievement (*18*, *19*, *26*). In other

STEM content areas, students' self-concept studies have been linked with their motivation (*46*, *47*). Self-efficacy has similarly been shown to positively correlate with student achievement in chemistry (*42*). Students' attitude toward learning chemistry is a metric that can be reliably measured, validated, and used to predict achievement in chemistry, thereby affording a temporal window of opportunity for timely academic intervention (*22*, *33*, *48*). Generally, researchers find that students with a more positive affective domain outperform students with a more negative affective domain (*27*, *32*, *33*, *45*). Therefore, chemistry instructors without access to the unique hidden constructs of their students' affective domain are losing a critical potential ability to fully serve their students' learning needs.

However, while research is increasing into students' nuanced affective domain in science education generally, and in chemistry education specifically, this work seems to bypass the two-year college student body and focuses instead at the secondary education and four-year college levels (*27*, *49–52*). Students in high school and at two-year and four-year institutions can be very different and therefore are expected to have measurably different affective characteristics. Funding, location, entrance policies, reputation are all potential influencers of attitudes (*53*). Indeed, studies have shown that students from comparable institutions can interpret and respond to attitude measuring instruments in slightly different ways (*49*, *50*, *54*, *55*).

Modeling Students' Affective Domain

Modeling data is an essential practice in modern scientific research as it offers a window into data comprehension and causation. Establishing testable theories of data causation affords the researcher the opportunity to manipulate the outcome in an intended direction. This is particularly powerful and relevant to an educational setting wherein an instructor may wish to test a teaching methodology, or provide a remedial academic intervention (*22*, *33*, *48*, *56*).

Modeling data requires assigning numbers to represent measurable quantities. If the variable of interest is directly measureable, say the height of an object, then assigning numbers to the object is relatively easy (e.g. a ruler). However, if the variable to be measured is an abstract concept that is not directly measurable (covert), such as self-concept or confidence, then assigning relevant numbers is much more difficult, and the abstract object is termed a "latent variable" or "factor".

Factors are themselves not directly measurable and might be mistaken to be only qualitative concepts. Examples of factors include the aforementioned sub-constructs of students' affective domains–self-efficacy, self-concept, and attitudes–and perhaps the more relatable concept of phobias. However, factors are quantitative and can be measured indirectly by identifying and quantifying their antecedent triggers (*57*). For example, a person's phobia of snakes can be triggered when asked questions about snakes, or asked to view images of snakes. Carefully constructed surveys can then be administered to quantify the person's phobia of snakes by directly quantifying their responses to the survey questions that triggered the phobia. Analysis of how these antecedent triggers–items, or manifest variables–manifest into the phobia itself is the objective of exploratory factor analysis (EFA) (*57*).

Assessing students' affective domain with EFA is a potentially profitable education strategy at all levels of chemistry education since it affords the educator insights into the covert factors that influence students' learning (*58*). It is advisable to follow EFA with confirmatory factor analysis (CFA). CFA is employed by researchers wanting to test their proposed relationships between manifest and latent variables (measurement model), and between latent variables (structural model). In practice, if one data set is used for EFA to explore hypothetical measurement and structural models, a second data set is used to confirm the proposed models using CFA. Research studies are now reporting CFA data in addition to EFA data more frequently (*32*, *51*, *54–56*, *59–61*). CFA is a useful statistical tool in applied research due to its ability to demonstrate causality in proposed measurement and structural models (models) (*62–64*). CFA is frequently performed using covariance based structural equation modeling (CB-SEM). However, there are alternatives to CB-SEM that are better suited to different circumstances.

As with CB-SEM, partial least squares structural equation modeling (PLS-SEM) allows latent variables to be indirectly measured (*65–68*). Whereas CB-SEM results include accepted measures of fit index that are used to critique the proposed models, no comparable measures of fit index exist for models produced from PLS-SEM (*64*, *67*, *69–73*). Instead, PLS-SEM models are assessed in terms of the significance of their path coefficients (β), data variance explanatory adjusted R^2 values, effect sizes (f^2 and q^2), and the predictive relevance values (Q^2) (*68*, *74*, *75*).

Modeling Students' Academic Self-Concept

Current research posits that students' academic self-concept is a multifaceted, hierarchical construct (*76*). Here, we are distinguishing academic self-concept from non-academic self-concept, both of which fall under the apex of global self-concept (*10*, *14*). Following decades of controversial findings (*10*, *76–79*), and pioneering work by Marsh (*80*) and Shavelson (*10*), the preferred hierarchical model of students' academic self-concept is aptly named the Marsh/Shavelson model (*81*). The Marsh/Shavelson model of academic self-concept consists of a hierarchy of discrete *math* and *verbal* self-concepts linking with specific subject area self-concepts (Figure 1) (*82*). An interesting aspect of the Marsh/Shavelson model, highlighted in Figure 1, is the apparent lack of significant interaction between the *math* and *verbal* academic self-concepts. This is a robust finding, and contrasts against the equally reproducible finding of a strong correlation between math and English achievement. This contrasting relationship between mathematics and verbal achievement and self-concept is visualized in the internal/external frame of reference (I/E) model of academic self-concept (*80*).

The I/E model of academic self-concept posits that students formulate their academic self-concepts under the influence of self-evaluation with other peers (*80*). The external self-evaluation occurs when students compare their own abilities against the abilities of other students in the same subject area, whereas the internal self-evaluation arises when students compare their own ability in a content area relative to their ability in another content area, for example, a comparison of academic ability in a mathematics versus a chemistry class (Figure 2) (*83*).

According to the I/E frame of reference model in Figure 2, students' academic achievement causes their academic self-concept, as indicated by the high positive paths (β ++) and low negative paths (β –). However, it does not dispute the fact that it is widely accepted, and backed by research, that academic achievement and academic self-concept are both cause and effect of each other (*84–87*). Figure 2 shows that mathematics and verbal achievement are typically strongly and positively correlated (R ++), whereas *mathematics* and *verbal* self-concept are essentially uncorrelated (R 0) (*80*). This latter finding indicates that students apparently do not perceive the interconnectedness of mathematics and English. Mathematics and science achievement are likewise positively correlated (R ++). In contrast to *mathematics* and *verbal* self-concept, *mathematics* and *science* self-concept are positively correlated (R ++) (*83*). A student with a high self-concept in mathematics will typically have a high self-concept in science due to the students' perceived interconnectedness of mathematics and science.

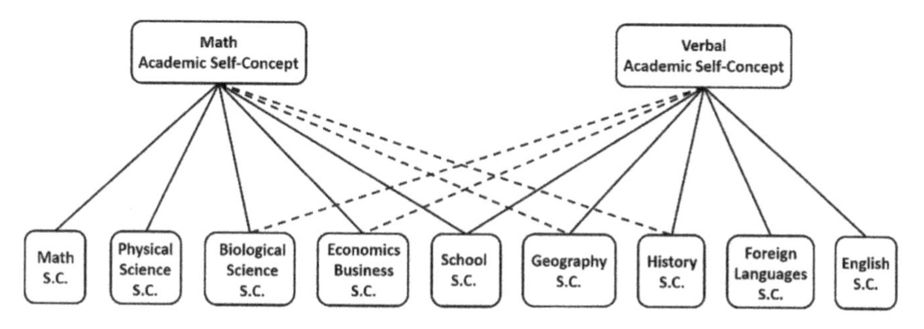

Figure 1. Academic self-concept structure of the Marsh/Shavelson model. Reproduced with permission from reference (81). Copyright 1988 The American Psychological Association.

While the I/E frame of reference model incorporates both students' internal and external frame of reference, a related self-concept construct known as the big-fish-little-pond effect (BFLPE) focuses solely on the students external frame of reference, and is unique to students' academic self-concept (*88–90*). The origin of the BFLPE stems from students' tendancy to compare their own academic achievements with their peers' academic achievements. The BFLPE is typically a negative effect of average group achievement on an individual student's academic self-concept. For example, a student will naturally compare his/her academic achievement in a course with the achievement of other students. According to the BFLPE, students of class-average ability will downplay their academic self-concept in a subject area if they perceive the class-average ability to be high. In reverse, students of average class ability will overstate their academic self-concept in a content area if they perceive the class-average ability to be low. Due to the open-door policy, students in two-year colleges possibly have an inflated academic self-concept.

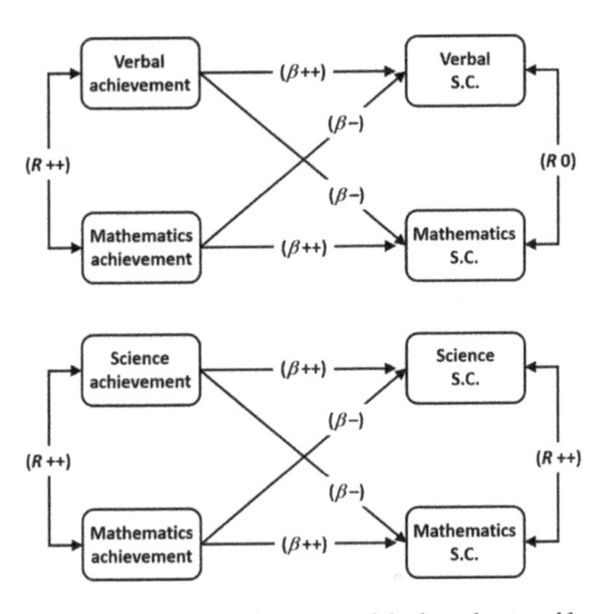

Figure 2. The I/E frame of reference model of academic self-concept.

Although the BFLPE has both negative and positive effects on students' academic self-concept, the negative effect is typically much stronger than the positive effect, rendering a net negative impact of the BFLPE on students' academic self-concept (*90*). The undeniably inflated value placed on premature academic advancement and its prevalence and coveted prestige for gifted students therefore could have a sinister impact on students' academic self-concept.

Present Study

There is a need for research into the affective domain of students in two-year colleges. The objective of the present study was to probe a known marker of student learning, namely self-concept; this was done during a 5-week Winter intersession with an introductory chemistry cohort. We chose to investigate the self-concept component of our students' affective domain using the Chemistry Self-Concept Inventory (CSCI) survey instrument (*26*). The CSCI survey was chosen from numerous other possible instruments designed to survey students' affective domain in chemistry due to its pedigree (*42*, *91–94*).

The CSCI instrument was developed from the Self Description Questionnaire III (SDQIII) (*95*) and has been shown to provide reliable and valid data (*79*, *96*). The CSCI instrument contains 40 statements compartmentalized into 5 subscales to represent and capture 5 unique latent variable components of self-concept relevant to learning chemistry: *mathematics*, *chemistry*, *academic*, *academic enjoyment,* and *creativity* self-concepts, respectively. Examples of *mathematics* self-concept statements include, "I am quite good at math," and "Math makes me feel inadequate." Students respond to each statement on a 7-point scale depending on how accurately the statement reflects their own viewpoint (1 = very inaccurate; 7 = very accurate). *Chemistry* self-concept statements are similar to *mathematics* self-concept statements but with the emphasis switched from mathematics to chemistry. Examples of *academic* and *academic enjoyment* self-concept statements are, "I get good marks in most academic subjects," and "I like most academic subjects," respectively. An example of a *creativity* self-concept statement is, "I am an imaginative person." The CSCI survey is available in the supplemental material of reference (*26*).

We assumed that our students' self-concept data would adhere to the Marsh/Shavelson model of self-concept (Figure 1). Due to its preferential applicability to research adhering to known theoretical models, coupled with its suitability to smaller sample sizes, PLS-SEM was adopted for the modeling of our students' self-concept data. PLS-SEM is suited to causal structural equation modeling when data is being modeled against a known theoretical model and when sample sizes are too small for reliable structural modeling with CB-SEM (*67*). PLS-SEM can often furnish robust statistical results from smaller sample sizes and is well placed as a modeling tool for exploratory research analysis in cases where the aim is to apply and explore an existing theoretical structural model (*97–99*). Herein, we report the results and implications following the measurement of students' self-concept during an introductory chemistry course at a large urban two-year college.

Research Questions

The specific research questions being asked in this study are:

1. Does the CSCI survey instrument, originally developed and administered in a traditional four-year institution serving mainly non-minority students, capture the same latent self-concept constructs when offered to the students at an urban two-year college, the majority of whom are minority students?
2. Can PLS-SEM be used to meaningfully represent our students' self-concept data according to the Marsh/Shavelson model?
3. Is there any evidence of the BFLPE in our students' self-concept data?

Methods

Students' self-concept was measured using Bauer's CSCI survey instrument (*26*). The CSCI survey consists of 40-items that probe 5 unique areas of self-concept, namely *mathematics*, *chemistry*, *academic*, *academic enjoyment*, and *creativity* self-concepts. CSCI item 3 and 15 were automatically excluded from the data acquisition, following Bauer's recommendation that these items are poor representatives of their intended factors (*26*). Therefore, students' responses to only 38 items from the CSCI instrument were initially analyzed. The CSCI survey, developed from the SDQIII survey, requires participants to respond to statements on a 7-point scale depending on whether the statement is a "very accurate" (scale = 7) or "very inaccurate" (scale = 1) representation of the students' own viewpoint (*95*). The CSCI survey probes aspects of learning that will directly or indirectly influence students' self-concept in chemistry.

Participants

The CSCI instrument was administered in paper format to 118 students enrolled during the Winter 2016 intersession in 6 separate introductory chemistry classes at a large two-year institution in southern California. Each class was led by a different instructor, so the teaching style and general learning approach of 6 separate instructors was potentially incorporated into our students' data. The number of useable surveys was 102 from week 1 and 84 from week 5. To be considered useable, surveys had to be completed by the same student twice during week 1 and again during week 5. The 16 surveys not useable in week 1 were due to the same student completing a survey only once during week 1. A further loss of 18 more surveys by week 5 resulted from a combination of student attrition and students adding the class after missing one of the surveys administered during week 1. Students who completed week 1 surveys were 50.0% female and 50.0% male. Among these students, 71.6%, 18.6%, and 9.8% self-identified as Hispanic/Latino, Asian/Pacific Islander, and other, respectively.

Data Collection and Analysis

CSCI survey responses were collected from 102 students who completed the survey twice during the first week of class. Administering the CSCI survey twice during the first week of class is important as it enables the measurement of test-retest reliability, which is important if students' self-reported responses are to be believed (*100*, *101*). A third CSCI survey was administered during week 5 to capture any changes in students' self-concept data occurring during the course.

Students' survey responses were manually input into Excel spreadsheets provided by Bauer (*26*). Since the CSCI is a self-reporting survey, which seeks to measure latent constructs that are sensitive to interpretation and are influenced by cultural norms (*82*) and frame of reference (*80*), we measured the handling of the CSCI instrument with our introductory chemistry students in terms of the internal consistency of items within each subscale (Cronbach's alpha) and in terms of test-retest reliability of students' self-reporting (Pearson correlation). Students' self-concept typically establishes and settles with time in a course (*18*, *27*). For this reason, students' CSCI data from week 5 was expected to be superior to week 1 data, and was preferentially inspected via EFA and PLS-SEM.

EFA of the internal structure of the CSCI instrument was performed using XLSTAT 2017.6 software and adopted the same principal components analysis with varimax rotation as found in the original work (*26*). The number of factors to extract from EFA was decided based on the eigenvalue-greater-than-one rule, coupled with manual inspection of the scree plot, and careful comparisons of our students' data against the Marsh/Shavelson self-concept model and Bauer's original work (*26*, *80*).

Results and Discussion

Descriptive Statistics

Out of Bauer's recommended 38 CSCI items, the four items composing the *creativity* self-concept construct were eliminated from further consideration since the *creativity* self-concept does not feature in the Marsh/Shavelson model (Figure 1). Other items were eliminated due to relatively low factor loadings. In all, 10 items were dropped, leaving only 28 CSCI items for further analysis. Dropping items from the analysis in this way is common practice during factor analysis, and is acceptable since the latent factor which the dropped items represent remains, and is still represented by the remaining items. Average scores for the 28 CSCI survey items ranged from 4.262 to 5.655, with standard deviations ranging from 1.297 to 1.909, for week 5 data (1-7 range, 4 midpoint). Higher item scores indicate more positive self-concepts (Table 1). Skewness values ranged from -0.847 to +0.801, and kurtosis values ranged from -1.004 to +1.120, for students' CSCI data, meaning that the data was normally distributed and suitable for broad statistical analysis.

Table 1. Descriptive Statistics and Reliability of Self-Concept Data from the CSCI Instrument[a]

Construct	*Item Number*	*Cronbachs Alpha*	*Composite Reliability*	*Mean (SD)*
Acad E	2	0.869	0.905	5.250 (1.388)
	6*			5.369 (1.564)
	10			5.321 (1.416)
	22*			5.286 (1.593)
	30*			5.405 (1.582)
	38*			5.655 (1.332)
Acad	18	0.817	0.873	5.131 (1.352)
	23			5.238 (1.306)
	26			5.095 (1.297)
	34			5.012 (1.367)
	39			4.893 (1.389)
Math	1	0.908	0.925	5.024 (1.655)
	5*			4.893 (1.909)
	9			4.714 (1.783)
	13*			4.952 (1.654)
	17			4.786 (1.582)

Continued on next page.

Table 1. (Continued). Descriptive Statistics and Reliability of Self-Concept Data from the CSCI Instrument[a]

Construct	*Item Number*	*Cronbachs Alpha*	*Composite Reliability*	*Mean (SD)*
	19*			4.869 (1.510)
	21*			5.048 (1.690)
	25			4.893 (1.581)
	29*			5.012 (1.476)
	37*			4.750 (1.779)
Chem	4*	0.893	0.918	4.762 (1.637)
	16			4.262 (1.381)
	20*			4.512 (1.783)
	24			4.333 (1.458)
	28*			4.429 (1.748)
	32*			4.512 (1.531)
	40*			5.060 (1.442)

[a] Week 5 ($N = 84$). * Negatively stated item is reversed before averaging.

It is important to verify the initial stability of students' self-reporting for each factor. A common way to do this is to measure the correlations for each factor between an initial survey and a second survey administered soon thereafter (test-retest). The results from the test-retest correlation of each factor in the CSCI instrument during week 1 of class were significant and ranged from $r = 0.56$ to $r = 0.67$ ($N = 102$, $p < 0.00001$) (*26, 27*). These correlation values suggest that our students' self-reporting was stable and therefore should be believed.

The suitability of our students' CSCI data for factor analysis was determined prior to performing EFA. The Kaiser-Meyer-Olkin (KMO) test of sampling adequacy measures the proportion of variance among survey items that might be due to a common variance (*102*). Lower proportions of variance are more suitable for FA, and a KMO value is calculated and reported as a value between 0 and 1 (KMO is inversely proportional to variance), with values > 0.70 indicating an acceptable KMO value for FA and structure detection. The KMO value for the 38-item CSCI instrument was 0.817, indicating the suitability of our students' data for EFA (*102*). Bartlett's test of sphericity, which addresses the hypothesis that the data submitted for EFA is in fact unrelated and further analysis for detecting structure is pointless, was also conducted (*103*). A significant Bartlett's test result (χ^2(df), $p < 0.05$) would confirm the existence of correlated data suitable for EFA. Bartlett's test of sphericity ind ($\chi^2(740) = 2941.903$, $p < 0.0001$). Together, the KMO and Bartlett's pre-factor analysis data posit that our students' CSCI survey data was suitable for EFA and structure detection.

Internal Consistency Reliability for CSCI

It is important when using an instrument designed to measure certain latent constructs to verify that it is indeed measuring those intended latent constructs, so that the data can be believed. The internal consistency reliability of a latent construct quantifies how well each item is representative of its intended latent construct. Measures of internal consistency reliability for each latent constructs–Cronbach's alpha and composite reliability–both of which range from 0.0 to 1.0, were in excess of the threshold minimum acceptable value of 0.70. Composite reliability tends to overestimate internal consistency reliability, whereas Cronbach's alpha is more conservative, and tends to underestimate the reliability of a construct. Therefore, a better estimate of internal consistency reliability is an average of these two measurements (*104*, *105*). Cronbach's alpha values ranged from 0.817 to 0.908, while composite reliability values ranged from 0.873 to 0.925, signaling that all items within the presented latent constructs of CSCI both belonged to, and were acceptably representative of, their latent contructs (Table 1).

Students' Self-Concept Data

PLS-SEM analysis of our students' data was performed on structural models that adhere to the Marsh/Shavelson model of self-concept (*81*). It is important when considering PLS structural models to be guided by known theory and not merely pursue patterns in students' data (*64*). Therefore, the *creativity* self-concept construct of CSCI was not considered further since it could not be placed into the Marsh/Shavelson model (Figure 1). Only CSCI survey items resulting in constructs of the Marsh/Shavelson model of self-concept are reported further.

Average construct scores were 5.381, 5.074, 4.894, and 4.553 for *academic enjoyment*, *academic*, *mathematics*, and *chemistry* self-concepts, respectively (Table 2). Our students' average *chemistry* self-concept score was higher than the same construct score observed by others in four-year institutions. Bauer recorded an average *chemistry* self-concept score of 3.88 (*26*). Lewis et al. reported an average *chemistry* self-concept score of 4.03 (*19*). We have insufficient data to claim that our students' *chemistry* self-concept data, with a value of 4.553, is statistically higher than the values reported in four-year institutions, and some urge against doing so (*26*). However, we note that it is higher, and speculate that if later confirmed to be statistically significant that this could be evidence of the BFLPE occurring with our students (*88*).

Average item scores were the lowest for the *chemistry* self-concept construct, in line with others' findings (Table 2). Notably, all self-concept scores were above the midpoint of 4, signaling that our students' academic self-concepts are generally positive. As expected from literature reports (*19*, *26*), item scores were on average highest for the *academic* and *academic enjoyment* self-concept constructs, probably indicating that these constructs were the driving forces behind the students' self-concept data overall, in line with the Marsh/Shavelson model (*81*). Bauer and Lewis et al. also recorded highest average self-concept scores for the *academic enjoyment* and *academic* constructs (Table 2) (*19, 26*).

Table 2. Average Construct Scores of Our Work Compared with the Literature[a]

Study	*Acad E*	*Acad*	*Math*	*Chem*
Our data	5.381	5.074	4.894	4.553
Bauer (*26*)	5.38	5.08	4.48	3.88
Lewis et al. (*19*)	5.24	5.02	5.44	4.03

[a] Week 5 ($N = 84$). All scores out of a possible range of 1 (low) to 7 (high).

Structural Model

Nielsen and Yezierski note that students' self-concept data is more accurately rendered later in a course than at the beginning (*27*). For this reason, our students' data taken during week 5 was considered the best representation of self-concept data and was used exclusively for structural equation modeling. PLS-SEM is a structural modeling technique that relies upon adherence to a known theoretical model. Therefore, PLS-SEM was performed on students' data from week 5, aligned to the Marsh/Shavelson model of self-concept (Figure 3). Ovals represent latent constructs of self-concept and those receiving path inputs display their adjusted R^2 values (explained variance). Arrows connecting latent constructs show standardized path coefficients (β, $*p < 0.05$; $**p < 0.001$). Rectangles show the manifest variables (survey items).

Relatively high adjusted R^2 values are a hallmark of a good PLS structural model (*68*). Adjusted R^2 values of 0.75, 0.50, or 0.25 for latent constructs are considered substantial, moderate, or weak, respectively. We selected the structural model shown in Figure 3 as the best model of our students' CSCI data, compared with other competing models, owing to its greater overall adjusted R^2 values, β values and f^2 effect sizes (Table 3) (*68*).

The structural equation model shown in Figure 3 contains 2 moderate and 1 weak adjusted R^2 variable and accounted for approximately 55% of the variance in students' *chemistry* self-concept data (R^2 x 100). The structural model also contains 2 large, 1 moderate, and 2 small f^2 effect sizes. Values of f^2 of 0.02, 0.15, and 0.35 are considered to be small, medium, and large effects, respectively (Table 3). The f^2 effect sizes relate to the adjusted R^2 values of the model. In general, effect sizes are used to communicate the relative influence and relationship strength between linked variables, which supplements other statistical information regarding their statistical significance (*68*). PLS-SEM has no accepted global measure of model fit. However, measures of predictive relevance (Q^2) and its corresponding effect size (q^2) have been used to evaluate structural models (*106*).

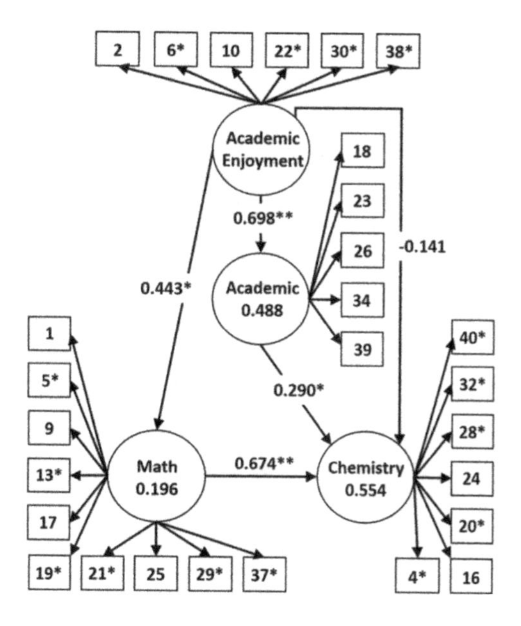

Figure 3. PLS-SEM model taken from the CSCI instrument administered during week 5 (N = 84).

Table 3. Structural Model Evaluation (f^2 Effect Size)[a]

Construct	*Adjusted R^2*	f^2		
		Acad E	*Acad*	*Math*
Acad	0.488	0.951		
Math	0.196	0.244		
Chem	0.554	0.022	0.098	0.830

[a] Week 5 ($N = 84$). Values for f^2 of 0.02, 0.15 and 0.35 are viewed as small, medium and large effects, respectively.

Blindfolding is a procedure used to evaluate the predictive relevance of PLS structural model, for example, how well the structural model represents the data. Blindfolding was performed on the structural model shown in Figure 3 to measure its cross-validated redundancy (Q^2) values (*68*). Latent constructs exhibiting a Q^2 value greater than 0 support the predictive relevance and validity of the structural model. Latent constructs with a Q^2 value below 0 build a model that lacks predictive relevance and validity. The structural model shown in Figure 3 has Q^2 values greater than 0 for all self-concept constructs with 1 large and 4 small q^2 effect sizes, and is therefore an acceptable representation of our students' self-concept data (Table 4). Values for q^2 of 0.02, 0.15, and 0.35 are considered to be small, medium, and large effect sizes, respectively (Table 4). The structural model shown above was chosen to represent our students' self-concept data because of its alignment with theory and because of its superior adjusted R^2,

f^2, Q^2, q^2, and β values (*68*). The integrity of the presented model was further investigated according to its reliability and validity (*101*).

Table 4. Structural Model Evaluation (q^2 Effect Size)[a]

Construct	Q^2	q^2			
		Acad E	*Acad*	*Math*	*Chem*
Acad	0.238			-0.063	-0.071
Math	0.108		-0.002		-0.024
Chem	0.345	0.000	0.037	0.325	

[a] Week 5 ($N = 84$). $Q^2 > 0$ indicates a model has predictive relevance for a latent construct; $Q^2 < 0$ indicates a lack of predictive relevance. Values for q^2 of 0.02, 0.15, and 0.35 are viewed as small, medium, and large effects, respectively.

Reliability

In addition to the favorable composite reliability measurements presented in Table 1, the model was probed in terms of the representational reliability of each manifest variable for its latent construct. Manifest variables function as antecedents of specific latent constructs, and as such should be more representative of these specific latent constructs than any others. The representational reliability of a manifest variable for its intended latent construct is gauged by its loading, which should be greater than loadings onto any unintended factors, and ideally should be > 0.70 on a scale of 0.0 to 1.0. All manifest variables loaded onto their intended factors and displayed average loadings of 0.774, 0.758, 0.740, and 0.780 for *academic enjoyment*, *academic*, *mathematics*, and *chemistry* self-concepts, respectively. The manifest variable reliability scores (loading squared) were on average above the minimum value of 0.50 (loading > 0.707), again pointing to meaningful and reliable data (*68*).

Validity

The Fornell-Larcker criterion (FLC) for discriminant validity was inspected. The FLC is used to address and prevent multicollinearity issues which can lead to systematic errors in models (*107*). The FLC is a common way to evaluate the degree of shared variance between latent constructs in a structural model and screen for multicollinearity issues. The FLC is adhered to when the average variance extracted (AVE) value of each latent construct exceeds the constructs' highest squared correlation with other latent constructs (*108*). It can be seen in Table 5 that all AVE values were higher than the squared correlations with other latent constructs. This result indicates that discriminant validity was achieved for the model, meaning that items belong to their designated constructs, do not belong elsewhere, and are being interpreted correctly.

Table 5. Discriminant Validity (Fornell-Larcker Criterion)[a]

	Acad E	*Acad*	*Math*	*Chem*	*AVE*
Acad E	1.000				0.605
Acad	0.488	1.000			0.580
Math	0.196	0.142	1.000		0.553
Chem	0.129	0.198	0.520	1.000	0.611

[a] Week 5 ($N = 84$). Cross-correlations shown for each construct.

Interpretation of the Structural Model

The significant path between the *mathematics* and *chemistry* self-concept constructs adheres to the internal/external frame of reference (I/E) model and can be attributed to our students' internal self-evaluation of academic content areas (Table 3) (*80, 109*). The structural model shows an insignificant negative path value connecting the *academic enjoyment* and *chemistry* self-concepts. This path is insignificant and so could have no meaning. Alternatively, this could suggest that our students do not find chemistry to be enjoyable. Although unfortunate if correct, perhaps this last statement is to be expected from a cohort of introductory chemistry students surveyed during an intensive 5-week Winter intersession course. The observation that *academic enjoyment* and *mathematics* self-concepts connect with a positive and significant β value, whereas *academic enjoyment* and *chemistry* self-concepts connect with a negative and insignificant β value, suggests that our students' outlook on mathematics and chemistry courses is quite different (*110*).

A possible explanation is that our students have experienced and enjoyed the reciprocity between effort and achievement in mathematics, and hence find mathematics to be an enjoyable, conquerable academic pursuit, whereas they have a pior negative experience in chemistry, whereby effort has not led to achievement. The students in this study have already had several mathematics classes before taking this chemistry class, so this speculative interpretation is possible. However, the students are (we assume) taking chemistry for the first time in college, so it is less certain that the given hypothesis to account for the students' probable malcontent with chemistry is probable. In lieu of their own personal academic experience, the students could be speculating based on a stereotypical viewpoint that chemistry is abstract and elusive. Of course, the path between *academic enjoyment* and *chemistry* self-concepts could just be a null. However, deletion of this path had a deleterious effect on the structural model, which is grounds for retaining it in the model presented here (*68*). The reason behind our students' different outlook on *mathematics* and *chemistry* self-concepts is not known by the authors at the time of writing but will be explored in future work.

Another aspect of the I/E frame of reference model, not shown in the structural model presented here, but found in a parallel analysis, is the significant and positive correlation between mathematics prerequisite grade and proceeding introductory chemistry grade ($r = 0.54$, $p < 0.001$, $N = 195$). This parallel analysis of the mathematics prerequisite was limited to a cohort of introductory chemistry students who completed the mathematics prerequisite course in a major term (Fall or Spring) and then completed introductory chemistry in the following major term (Fall or Spring). Encouragingly, the significant path connecting the *mathematics* and *chemistry* self-concepts (Figure 3), and the correlating mathematics and chemistry achievement data, likely means that our students value their mathematics prerequisite requirement, and they believe it is an important predictor of future success in introductory chemistry. While it is important for students to understand the reason for assigning a prerequisite to a class, it is arguably equally important for students and instructors alike to be made aware that the common mathematics prerequisite is not the only potential obstacle to success in introductory chemistry.

Implications for Prerequisites and Instruction

Whereas chemistry education research literature is rich with examples of non-mathematical prerequisites having a positive influence on chemistry education outcomes, curriculum development has been slow to adopt any of these research findings. This is particularly relevant and pressing for chemistry education at two-year colleges, which typically serve under-prepared students who require academic remediation to varying extents, and in today's educational climate which is generally increasing in the diversity of its student body (*53*).

The results presented herein complement the growing evidence in the chemistry education literature that current mathematical pre-requisite scores are probably insufficiently detailed for the role of informing students and instructors of students' preparedness for a particular chemistry class, or their likelihood of success during the chemistry class (*55*, *111*, *112*). Sensible additions to the existing mathematical prerequisites should include as concomitants attitudinal studies, wherein students and instructors engage in transparent and open-ended conversations about the role of students' self-concept, attitudes, and other relevant subcomponents of the affective domain.

Students' self-concept is a small but significant component of the learning process that should be adequately addressed as part of effective instruction, and incorporated into everyday planning and discourse within chemistry departments, at all levels, not just in the literature publications and conference proceedings of a few select chemistry groups. Students' self-concept is an area of their affective domain that should be brought into their awareness and into the open at the faculty and administrative levels, including discussions on student learning outcomes and curriculum planning.

Conclusions

In response to our research questions, the implementation of Bauer's CSCI survey instrument can provide a meaningful analysis and interpretation of students' self-concept data at the two-year college level, as evidenced by average construct values comparable to literature reports (Table 2) (*19*, *26*), and by the structural model that is rooted in the Marsh/Shavelson frame of reference model of self-concept (Figure 3). This study demonstrates the use of PLS-SEM to successfully model students' self-concept data in accordance with the Marsh/Shavelson model of self-concepts, and shows that meaningful data can be interpreted after only five weeks of an introductory chemistry course. A regular semester consists of 3 five-week intervals suggesting that students' self-concepts could potentially be monitored on three separate occasions during a semester to analyze possible changes in individual students' self-concepts. This study also shows a possible BFLPE for our students' *chemistry* self-concept, although further research is necessary to confirm this hypothesis.

PLS-SEM offers chemistry instructors in two-year and four-year institutions the opportunity to explore the covert constructs that can impact students' learning, and which serve to differentiate their learning needs. These covert constructs include self-concept and other components of students' affective domain. When adequately charted, students' affective domain can be navigated to cater the unique learning needs of each student in a classroom. PLS-SEM is well suited to the small cohort sizes often found in two-year colleges, where attitudinal research in chemistry is lacking and is very much needed. The authors suggest that measuring students' affective domain is a worthy action item as part of a comprehensive instructional plan in any introductory level chemistry class.

Following the observation in this work that our students are possibly not enjoying their introductory chemistry class, we plan to use PLS-SEM to monitor and evaluate the influence and efficacy of potential academic interventions designed to ameliorate our students' enjoyment of introductory chemistry. Extending this self-concept study to a regular semester-length class is currently underway in our department. A vertical study of students' self-concepts during their passage from introductory chemistry to general chemistry to organic chemistry would be beneficial to future departmental planning and curriculum design, and is currently in the planning stages in our department.

Acknowledgments

We would like to acknowledge financial support from the U.S. Department of Education HSI STEM, Jardin de STEM, P031C160250. We are also grateful to the East Los Angeles College Chemistry Clubs (POBC/MESA), and to the numerous chemistry professors who allowed us to survey their students.

References

1. Bloom, B. S.; Engelhart, M. D.; Hill, W. H.; Furst, E. J. *Taxonomy of Educational Objectives. Handbook I: Cognitive Domain*; David McKay Company, Inc.: New York, 1956.
2. Krathwohl, D. R.; Bloom, B. S.; Masia, B. B. *Taxonomy of Educational Objectives. Handbook II: Affective Domain*; David McKay Company, Inc.: New York, 1964.

3. Bunce, D. M.; Hutchinson, K. D. The use of the GALT (Group Assessment of Logical Thinking) as a predictor of academic success in college chemistry. *J. Chem. Educ.* **1993**, *70*, 183–187.
4. Spencer, H. E. Mathematical SAT test scores and college chemistry grades. *J. Chem. Educ.* **1996**, *73*, 1150–1153.
5. Wagner, E. P.; Sasser, H.; DiBiase, W. J. Predicting students at risk in general chemistry using pre-semester assessments and demographic information. *J. Chem. Educ.* **2002**, *79*, 749–755.
6. Ewing, M.; Huff, K.; Andrews, M.; King, K. *Assessing the Reliability of Skills Measured by the SAT*; Research Notes RN-24; The College Board: New York, December 2005; pp 1–8.
7. Lewis, S. E.; Lewis, J. E. Predicting at-risk students in general chemistry: Comparing formal thought to a general achievement measure. *Chem. Educ. Res. Pract.* **2007**, *8*, 32–51.
8. Paavola, S.; Lipponen, L.; Hakkarainen, K. Models of innovative knowledge communities and three metaphors of learning. *Rev. Educ. Res.* **2004**, *74*, 557–576.
9. Rosenberg, M. *The Measurement of Self-Esteem: Society and the Adolescent Self-Image*; Princeton University Press: Princeton, NJ, 1965.
10. Shavelson, R. J.; Hubner, J. J.; Stanton, G. C. Self-concept: Validation of construct interpretations. *Rev. Educ. Res.* **1976**, *46*, 407–441.
11. Bandura, A. Self-efficacy: Toward a unifying theory of behavioral change. *Psychol. Rev.* **1977**, *84*, 191–215.
12. Marsh, H. W. Self-concept: The application of a frame of reference model to explain paradoxical results. *Aust. J. Educ.* **1984**, *28*, 165–181.
13. Haertel, G. D.; Walberg, H. J.; Weinstein, T. Psychological models of educational performance: A theoretical synthesis of constructs. *Rev. Educ. Res.* **1983**, *53*, 75–91.
14. Byrne, B. M. The general/academic self-concept nomologial network: A review of construct validation research. *Rev. Educ. Res.* **1984**, *54*, 427–456.
15. McLeod, D. B. In *Research on Affect in Mathematics Education: A Reconceptualization*; Grouws, D. A., Ed.; *Handbook of Research on Mathematics Teaching and Learning*; Macmillan Publishing Company: New York, 1992; pp 575–596.
16. Edmondson, K. M.; Novak, J. D. The interplay of scientific epistemological views, learning strategies, and attitudes of college students. *J. Res. Sci. Teach.* **1993**, *30*, 547–559.
17. Bong, M.; Skaalvik, E. M. Academic self-concept and self-efficacy: How different are they really? *Educ. Psychol. Rev.* **2003**, *15*, 1–40.
18. Nieswandt, M. Student affect and conceptual understanding in learning chemistry. *J. Res. Sci. Teach.* **2007**, *44*, 908–937.
19. Lewis, S. E.; Shaw, J. L.; Heitz, J. O.; Webster, G. H. Attitude counts: Self-concept and success in general chemistry. *J. Chem. Educ.* **2009**, *86*, 744–749.
20. Haladyna, T.; Shaughnessy, J. Attitudes toward science: A quantitative synthesis. *Sci. Educ.* **1982**, *66*, 547–563.

21. Shrigley, R. L.; Koballa, T. R.; Simpson, R. D. Defining attitude for science educators. *J. Res. Sci. Teach.* **1988**, *25*, 659–678.
22. Osborne, J.; Simon, S.; Collins, S. Attitudes towards science: A review of the literature and its implications. *Int. J. Sci. Educ.* **2003**, *25*, 1049–1079.
23. Shavelson, R. J.; Bolus, R. Self-concept: The interplay of theory and methods. *J. Educ. Psychol.* **1982**, *74*, 3–17.
24. Marsh, H. W.; Walker, R.; Debus, R. Subject-specific components of academic self-concept and self-efficacy. *Contemp. Educ. Psychol.* **1991**, *16*, 331–345.
25. Wilkins, J. L. M. Mathematics and science self-concept: An international investigation. *J. Exp. Educ.* **2004**, *72*, 331–346.
26. Bauer, C. F. Beyond "student attitudes": Chemistry self-concept inventory for assessment of the affective component of student learning. *J. Chem. Educ.* **2005**, *82*, 1864–1870.
27. Nielsen, S. E.; Yezierski, E. Exploring the structure and function of the chemistry self-concept inventory with high school chemistry students. *J. Chem. Educ.* **2015**, *92*, 1782–1789.
28. Bandura, A. *Self-Efficacy: The Exercise of Control*; W. H. Freeman and Company: New York, 1997.
29. Zimmerman, B. J. Academic studying and the development of personal skill: A self-regulatory perspective. *Educ. Psychol.* **1998**, *33*, 73–86.
30. Kan, A.; Akbas, A. Affective factors that influence chemistry achievement (attitude and self-efficacy) and the power of these factors to predict chemistry achievement–I. *J. Turk. Sci. Educ.* **2006**, *3*, 76–85.
31. Uzuntiryaki, E.; Aydin, Y. C. Development and validation of chemistry self-efficacy scale for college students. *Res. Sci. Educ.* **2009**, *39*, 539–551.
32. Ferrell, B.; Barbera, J. Analysis of students' self-efficacy, interest, and effort beliefs in general chemistry. *Chem. Educ. Res. Pract.* **2015**, *16*, 318–337.
33. Vishnumolakala, V. R.; Southam, D. C.; Treagust, D. F.; Mocerino, M.; Qureshi, S. Students' attitudes, self-efficacy and experiences in a modified process-oriented guided inquiry learning undergraduate chemistry classroom. *Chem. Educ. Res. Pract.* **2017**, *18*, 340–352.
34. Triandis, H. *Attitude and Attitude Change*; Wiley: New York, 1971.
35. Aiken, L. R. In *Recent Developments in Affective Measurement*; Payne, D. A., Ed.; Recent Developments in Affective Measurement; Jossey-Bass: San Francisco, CA, 1980; pp 1–24.
36. Mager, R. F. *Developing Attitude toward Learning*, 2nd ed.; Lake Publishing Co.: Belmont, CA, 1984.
37. Eagly, A. H.; Chaiken, S. In *Attitude Strength: Antecedents and Consequences*; Petty, R. E.; Krosnick, J. A., Ed.; Lawrence Erlbaum Associates: Mahwah, NJ, 1995; pp 413–432.
38. Cukrowska, E.; Staskun, M. G.; Schoeman, H. S. Attitudes towards chemistry and their relationship to student achievement in introductory chemistry courses. *S. Afr. Tydskr. Chem.* **1999**, *52*, 8–14.
39. Salta, K.; Tzougraki, C. Attitudes toward chemistry among 11th grade students in high schools in Greece. *Sci. Educ.* **2004**, *88*, 535–547.

40. Oskamp, S.; Schultz, P. W. *Attitudes and opinions,* 3rd ed.; Psychology Press: New York, 2005.
41. Grove, N.; Bretz, S. L. CHEMX: An instrument to assess students' cognitive expectations for learning chemistry. *J. Chem. Educ.* **2007**, *84*, 1524–1929.
42. Barbera, J.; Adams, W. K.; Wieman, C. E.; Perkins, K. K. Modifying and validating the Colorado Learning Attitudes about Science Survey for use in chemistry. *J. Chem. Educ.* **2008**, *85*, 1435–1439.
43. Cheung, D. Students' attitudes toward chemistry lessons: The interaction effect between grade level and gender. *Res. Sci. Educ.* **2009**, *39*, 75–91.
44. Heredia, K.; Lewis, J. E. A psychometric evaluation of the Colorado Learning Attitudes about Science Survey for use in chemistry. *J. Chem. Educ.* **2012**, *89*, 436–441.
45. Else-Quest, N. M.; Mineo, C. C.; Higgins, A. Math and science attitudes and achievement at the intersection of gender and ethnicity. *Psychol. Women Q.* **2013**, *37*, 293–309.
46. Bong, M.; Clark, R. E. Comparison between self-concept and self-efficacy in academic motivation research. *Educ. Psychol.* **1999**, *34*, 139–154.
47. Preckel, F.; Goetz, T.; Pekrun, R.; Kleine, M. Gender differences in gifted and average-ability students: Comparing girls' and boys' achievement, self-concept, interest, and motivation in mathematics. *Gift. Child Q.* **2008**, *52*, 146–159.
48. Cook, E.; Kennedy, E.; McGuire, S. Y. Effect of teaching metacognitive learning strategies on performance in general chemistry courses. *J. Chem. Educ.* **2013**, *90*, 961–967.
49. Brown, S. J.; Sharma, B. N.; Wakeling, L.; Naiker, M.; Chandra, S.; Gopalan, R. D.; Bilimoria, V. B. Quantifying attitude to chemistry in students at the University of the South Pacific. *Chem. Educ. Res. Pract.* **2014**, *15*, 184–191.
50. Kahveci, A. Assessing high school students' attitudes toward chemistry with a shortened semantic differential. *Chem. Educ. Res. Pract.* **2015**, *16*, 283–292.
51. Villafañe, S. M.; Lewis, J. E. Exploring a measure of science attitude for different groups of students enrolled in introductory college chemistry. *Chem. Educ. Res. Pract.* **2016**, *17*, 731–742.
52. Chan, J. Y. K.; Bauer, C. F. Learning and studying strategies used by general chemistry students with different affective characteristics. *Chem. Educ. Res. Pract.* **2016**, *17*, 675–684.
53. Malcom, S.; Feder, M. *Barriers and Opportunities for 2-Year and 4-Year STEM Degrees*; The National Academies Press: Washington, DC, 2016.
54. Xu, X.; Lewis, J. E. Refinement of a chemistry attitude measure for college students. *J. Chem. Educ.* **2011**, *88*, 561–568.
55. Xu, X.; Villafane, S. M.; Lewis, J. E. College students' attitudes toward chemistry, conceptual knowledge and achievement: Structural equation model analysis. *Chem. Educ. Res. Pract.* **2013**, *14*, 188–200.
56. Vishnumolakala, V. R.; Southam, D. C.; Treagust, D. F.; Mocerino, M. Latent constructs of the students' assessment of their learning gains instrument

following instruction in stereochemistry. *Chem. Educ. Res. Pract.* **2016**, *17*, 309–319.
57. Thompson, B. *Exploratory and Confirmatory Factor Analysis: Understanding Concepts and Applications*; American Psychological Association: Washington, DC, 2004.
58. Kahveci, M.; Orgill, M., Ed.; *Affective Dimensions in Chemistry Education*; Springer: Dordrecht, 2015.
59. Brandriet, A. R.; Ward, R. M.; Bretz, S. L. Modeling meaningful learning in chemistry using structural equation modeling. *Chem. Educ. Res. Pract.* **2013**, *14*, 421–430.
60. Ferrell, B.; Phillips, M. M.; Barbera, J. Connecting achievement motivation to performance in general chemistry. *Chem. Educ. Res. Pract.* **2016**, *17*, 1054–1066.
61. Liu, Y.; Ferrell, B.; Barbera, J.; Lewis, J. E. Development and evaluation of a chemistry-specific version of the academic motivation scale (AMS-Chemistry). *Chem. Educ. Res. Pract.* **2017**, *18*, 191–213.
62. Blalock, H. M. Multiple causation, indirect measurement and generalizability in the social sciences. *Synthese.* **1986**, *68*, 13–36.
63. Brown, T. A. *Confirmatory Factor Analysis for Applied Research*, 2nd ed.; The Guildford Press: New York, 2014.
64. Kline, R. B. *Principles and Practice of Structural Equation Modeling*, 4th ed.; The Guildford Press: New York, 2015.
65. Rigdon, E. E. Structural Equation Modeling. In *Modern Methods for Business Research*; Marcoulides, G. A., Ed.; Lawrence Erlbaum: Mahwah, NJ, 1998; pp 251–294.
66. Lohmöller J.-B. *Latent Variable Path Modeling with Partial Least Squares*; Physica: Heidelberg, Germany,1989.
67. Hair, J. F.; Ringle, C. M.; Sarstedt, M. PLS-SEM: Indeed a silver bullet. *J. Mark. Theory Pract.* **2011**, *19*, 139–151.
68. Hair, J. F.; Hult, G. T. M.; Ringle, C. M.; Sarstedt, M. *A Primer on Partial Least Squares Structural Equation Modeling (PLS-SEM)*; 2nd ed.; Sage Publications, Inc.: Thousand Oaks, CA, 2016.
69. Hu, L.; Bentler, P. M. Cutoff criteria for fit indexes in covariance structure analysis: Conventional criteria versus new alternatives. *Struct. Equ. Model.* **1999**, *6*, 1–55.
70. Schreiber, J. B.; Nora, A.; Stage, F. K.; Barlow, E. A.; King, J. Reporting structural equation modeling and confirmatory factor analysis results: A review. *J. Educ. Res.* **2006**, *99*, 323–337.
71. Schreiber, J. B. Core reporting practices in structural equation modeling. *Res. Social Adm. Pharm.* **2008**, *4*, 83–97.
72. Cudeck, R. Analysis of correlation matrices using covariance structure models. *Psychol. Bull.* **1989**, *105*, 317–327.
73. Chin, W. W. In *Handbook of Partial Least Squares: Concepts, Methods and Applications*; Handbooks of Computational Statistics; Vinzi, V. E.; Chin, W. W.; Henseler, J.; Wang, H., Ed.; Springer: New York,, 2010; pp 655–690.
74. Stone, M. Cross validatory choice and assessment of statistical predictions. *J. Royal Stat. Soc.* **1974**, *36*, 111–147.

75. Geisser, S. The predictive sample reuse method with applications. *J. Am. Stat. Assoc.* **1975**, *70*, 320–328.
76. Byrne, B. M.; Shavelson, R. J. On the structure of social self-concept for pre, early, and late adolescents: A test of the Shavelson et al. (1976) model. *J. Pers. Soc. Psychol.* **1996**, *70*, 599–613.
77. Wylie, R. C. *The Self-Concept: A Review of Methodological Considerations and Measuring Instruments*; University of Nebraska Press: Lincoln, NE, 1974.
78. Wylie, R. C. *The Self-Concept: Vol. 2. Theory and Research on Selected Topics*; University of Nebraska Press: Lincoln, NE, 1979.
79. Wylie, R. C. *Measures of Self-Concept*; University of Nebraska Press: Lincoln, NE, 1989.
80. Marsh, H. W. Verbal and math self-concepts: An internal/external frame of reference model. *Am. Educ. Res. J.* **1986**, *23*, 129–149.
81. Marsh, H. W.; Byrne, B. M.; Shavelson, R. J. A multifaceted academic self-concept: Its hierarchical structure and its relation to academic achievement. *J. Educ. Psychol.* **1988**, *80*, 366–380.
82. Byrne, B. M. Validating the measurement and structure of self-concept: Snapshots of past, present, and future research. *Am. Psychol.* **2002**, *57*, 897–909.
83. Chiu, M. S. Achievements and self-concepts in a comparison of math and science: Exploring the internal/external frame of reference model across 28 countries. *Educ. Res. Eval.* **2008**, *14*, 235–254.
84. Guay, F.; Marsh, H. W.; Boivin, M. Academic self-concept and academic achievement: Developmental perspectives on their causal ordering. *J. Educ. Psychol.* **2003**, *95*, 124–136.
85. Valentine, J. C.; DuBois, D. L.; Cooper, H. The relation between self-beliefs and academic achievement: A meta-analytic review. *Educ. Psychol.* **2004**, *39*, 111–333.
86. Marsh, H. W.; Chanal, J. P.; Sarrazin, P. G. Self-belief does make a difference: A reciprocal effects model of the causal ordering of physical self-concept and gymnastics performance. *J. Sports Sci.* **2006**, *24*, 101–111.
87. Marsh, H. W.; Craven, R. G. Reciprocal effects of self-concept and performance from a multidimensional perspective. *Perspect. Psychol. Sci.* **2006**, *1*, 133–163.
88. Marsh, H. W.; Parker, J. W. Determinants of student self-concept: Is it better to be a relatively large fish in a small pond even if you don't learn to swim as well? *J. Pers. Soc. Psychol.* **1984**, *47*, 213–231.
89. Marsh, H. W. The hierarchical structure of self-concept and the application of hierarchical confirmatory factor analysis. *J. Educ. Meas.* **1987**, *24*, 17–39.
90. Marsh, H. W.; Craven, R. G. In *Self-Concept Theory, Research, and Practice: Advances for the New Millennium*; Marsh, R. W.; Craven, R. G., Ed.; University of Western Australia: Sydney, New South Wales, Australia, 2000; pp 75–91.
91. Bauer, C. F. Attitude towards chemistry: A semantic differential instrument for assessing curriculum impacts. *J. Chem. Educ.* **2008**, *85*, 1440–1445.

92. Hockings, S. C.; DeAngelis, K. J.; Frey, R. F. Peer-led team learning in general chemistry: Implementation and evaluation. *J. Chem. Educ.* **2008**, *85*, 990–996.
93. Chatterjee, S.; Williamson, V. M.; McCann, K.; Peck, M. L. Surveying students' attitudes and perceptions toward guided-inquiry and open-inquiry laboratories. *J. Chem. Educ.* **2009**, *86*, 1427–1432.
94. Reardon, R. F.; Traverse, M. A.; Feakes, D. A.; Gibbs, K. A.; Rohde, R. E. Discovering the determinants of chemistry course perceptions in undergraduate students. *J. Chem. Educ.* **2010**, *87*, 643–646.
95. Marsh, H. W.; O'Neill, R. Self Description questionnaire III (SDQIII): The construct validity of multidimensional self-concept ratings by late adolescents. *J. Educ. Meas.* **1984**, *21*, 153–174.
96. Byrne, B. M.; Shavelson, R. J. On the structure of adolescent self-concept. *J. Educ. Psychol.* **1986**, *78*, 474–481.
97. Wold, H. In *Systems under Indirect Observations: Causality, Structure, Prediction. Part II*; Jöreskog, K. G.; Wold, H., Ed.; North-Holland: Amsterdam, 1982; pp 1–54.
98. Reinartz, W. J.; Haenlein, M.; Henseler, J. An empirical comparison of the efficacy of covariance-based and variance-based SEM. *Int. J. Res. Mark.* **2009**, *26*, 332–344.
99. Hair, J. F.; Sarstedt, M.; Hopkins, L.; Kuppelwieser, V. G. Partial least squares structural equation modeling (PLS-SEM): An emerging tool in business research. *Eur. Bus. Rev.* **2014**, *26*, 106–121.
100. *Standards for Educational and Psychological Testing*; American Educational Research Association, American Psychological Association, National Council on Measurement in Education: Washington, DC, 1999.
101. Arjoon, J. A.; Xu, X.; Lewis, J. E. Understanding the state of the art for measurement in chemistry education research: Examining the psychometric evidence. *J. Chem. Educ.* **2013**, *90*, 536–545.
102. Kaiser, H. F. Analysis of factorial simplicity. *Psychometrika* **1974**, *39*, 31–36.
103. Tobias, S.; Carlson, J. E. Brief report: Bartlett's test of sphericity and chance findings in factor analysis. *Multivariate Behav. Res.* **1969**, *4*, 375–377.
104. Davidshofer, C. O.; Murphy, K. R. *Psychological Testing: Principles and Testing*, 6th ed.; Pearson: Upper Sadler River, NJ, 2005.
105. Sijtsma, K. On the use, the misuse, and the very limited usefulness of Cronbach's alpha. *Psychometrika.* **2009**, *74*, 107–120.
106. Sharma, P. N.; Kim, K. H. In *Proceedings of the 33rd International Conference of Information Systems*; Orlando, FL, December 16–19, 2012; Huang, M.-H.; Piccoli, G.; Sambamurthy, V., Ed.; Association for Information Systems, AIS Electronic Library: Atlanta, GA, 2012.
107. Montgomery, D. C.; Peck, E. A. *Introduction to Linear Regression Analysis*; J. Wiley and Sons: New York, 1982.
108. Fornell, C.; Larcker, D. F. Evaluating structural equation models with unobservable variables and measurement error. *J. Mark. Res.* **1981**, *18*, 39–50.

109. Chiu, M.-S. The internal/external frame of reference model, big-fish-little-pond effect, and combined model for mathematics and science. *J. Educ. Psychol.* **2012**, *104*, 87–107.
110. Marsh, H. W.; Yeung, A. S. An extension of the internal/external frame of reference model: A response to Bong (1998). *Multivariate Behav. Res.* **2001**, *36*, 389–420.
111. Easter, D. C. Factors influencing student prerequisite preparation for and subsequent performance in college chemistry two: A statistical investigation. *J. Chem. Educ.* **2010**, *87*, 535–540.
112. Allenbaugh, R. J.; Herrera, K. M. Pre-assessment and peer tutoring as measures to improve performance in gateway general chemistry classes. *Chem. Educ. Res. Pract.* **2014**, *15*, 620–627.

Chapter 7

In-Class Worksheets for Student Engagement and Success

Gaumani Gyanwali*

Division of Math and Sciences, University of Arkansas Rich Mountain, 1100 College Drive, Mena, Arizona 71953, United States
***E-mail: gyanwali@uarichmountain.edu.**

In-class worksheets were designed and prepared by the instructor to support student learning and engagement in a general chemistry lecture class. In-class worksheets are helpful in raising student engagement by providing a space for keeping valuable information during lecture, helping students to be more focused, and solving calculation problems step-by-step. The worksheets helped to easily cover the course material while students recorded all key information. In-class worksheets are constantly updated in order to address the student need using feedback from student surveys. Students who regularly use these in-class worksheets have better course grades and a greater chance of retention and completion.

Introduction

During several years of teaching chemistry and other physical science courses, I was always looking for a way to address some issues students had while trying to conquer these math-intense courses. During the past few years, I have found that many students simply would not take notes during class and were likely to forget the content when they need it. Students that did take in-class notes were not organized enough to know where they took the notes and were stressed while trying to retrieve them. Students are more dependent on the internet and open web searches for crucial background information on specific content, and many come up with wrong information and misconceptions. Students may also understand the calculation steps and the reason for using a particular formula during the lecture time, but later while solving similar problems on assignments and exams, they lose that connection or formula. A majority of students in a class would like to know how a multi-step calculation problem is solved and want to keep that process safely in a place that can be easily found when needed.

Studies have found that in-class worksheets make students more active during class, increase participation, and are useful in reviewing by providing timely feedback (*1–3*). Another study found that students using pen and paper to take notes during lecture have advantages over students using laptops due to slower processing in keyboard writing (*4*). These in-class worksheets provide the opportunity for group discussions and an interactive problem-solving environment within the lecture sessions. They have also been found to improve undergraduate learning in biochemistry (*5*), biology (*6*) and physics (*7*).

I have utilized in-class worksheets as a simple and easily available technique to address issues related to student engagement, interaction, participation, active learning, and retention of concepts. I have not only used in-class worksheets to provide students with dependable resources but have also collected student suggestions and input in order to improve the process of learning. Constant learning through modification, improvements, and adjustments to student needs are also the qualities of a good teacher. Therefore, in-class worksheets are not simply "class handouts" but a major part of a student-centered approach with an ongoing and research-based adjustment of course contents and teaching methods.

What Are In-Class Worksheets?

As the name suggests, in-class worksheets are paper handouts given to students during a lecture. Students are required to complete the in-class work sheet within the scheduled class period. Several times during the lecture, many open-ended questions are asked, formulas written and stated, relationships between quantities analyzed, and mathematical problems solved by both the instructor and the students together. This brings a lively instructor-student interaction and opportunity for students to ask any questions they have. Students are expected to keep their completed worksheets in an organized binder. Very seldom are the worksheets graded or evaluated for points towards the course grade. If graded, participation points are allocated not to exceed 2.5% of the total course grade.

An in-class worksheet consists of two sections: chapter background, and problems and questions. The chapter background is designed to give students an overview for the content and concepts that will be covered in the chapter. In-class problems and questions are meant to discuss, analyze, and solve problems to cover the chapter content. The two sections are described below.

In-Class Worksheets: Chapter Background

The content for the chapter background may come from the previous chapter or from students' own understanding of the concept prior to the class. While discussing or answering these questions, an opportunity is provided for students to recall previous knowledge of the subject matter. Then, correct and logical (many times, straightforward) information is provided. The background discussion sessions provided by the in-class worksheets are helpful for both the students and the teacher in many ways (*8*). By completing the chapter background discussion, students will be able to:

1) Review the information already discussed in the previous chapter(s).

2) Confidently store the information they had or verify what they thought.

3) Build a foundation of knowledge for new or more challenging content discussed in the chapter.

4) Maintain alignment with fellow students before starting the chapter content.

5) Participate in class/group discussion and interact with the instructor.

An example of an in-class worksheet chapter background is given in Figure 1. This example is related to the Stoichiometry chapter in the General Chemistry I course. Only two questions are presented as a sample. Students actively take part in discussion with the instructor to answer these questions before starting the actual content of the chapter.

One of the crucial pieces of background information needed for successful understanding and foundation building in this chapter is the 'atomic mass' of elements. That is the reason why the first question (Question a. in Figure 1) asks students whether they have retained this information from the previous chapter. It also cautions them not to be confused with the other similar term 'mass number'. The other information that is important while solving stoichiometric problems is to know the molar ratio of molecules in a given balanced chemical equation (Question b. in Figure 1). If the chemical-to-chemical molar ratio is applied correctly while doing stoichiometric conversions, most problems are solved by avoiding the 'mistake traps'. Also in this example, the importance of correctly balanced equations covered in a previous chapter is connected carefully.

Chapter 5: Stoichiometry
Chapter Background

a. Where do you find atomic masses of elements? What unit is it expressed in? Is the atomic mass the same as mass number?
b. Find the number of moles of each species in the given equation:

$$2\ H_2O\ (l) \rightarrow 2\ H_2\ (g) + O_2\ (g)$$

Figure 1. Example of chapter background questions in an in-class worksheet for the stoichiometry chapter of General Chemistry I.

In-Class Worksheets: Problems and Questions

The major and crucial part of the in-class worksheet is the Problems and Questions section. After initial discussion of the chapter background, the rest of the class time is spent analyzing, calculating, and documenting what was learned. The content of this section comes directly or indirectly from lecture presentation slides. The focus is to practice crucial, and possibly difficult, mathematical calculations students must understand. In most cases, students will need the instructor's support to understand and apply the underlying mathematical concept. An example of the In-class Questions and Problems is given in Figure 2 below.

Chapter 5: Stoichiometry
In-Class Questions and Problems

1. What is one mole?
2. Avogadro's number =
3. Calculate the mass of one nitrogen atom.
4. Calculate the mass of one oxygen molecule.
5. How many molecules of methane are in 0.4 picogram of methane?

Figure 2. A sample of a portion of an in-class worksheet used in the stoichiometry chapter of General Chemistry I. Space is provided for student responses in an actual in-class worksheet.

The In-class Questions and Problems section is a mixture of qualitative and quantitative questions ranging from simple definition to multi-step mathematical calculations. It also includes any constant values students are responsible for knowing (Figure 2, Question #2). Also included are two similarly stated mathematical problems having different steps for the calculation due to a one-word difference ('atom' and 'molecule' in Figure 2, Questions #3 and #4). This section could include all question types including multiple-choice, true/false, filling in the blanks, matching, and writing formula and equations. With several of these carefully designed practice questions, the objective is to provide the students with an opportunity to interact while doing calculation problems in a lecture class and be confident in their recorded answers. In general, five types of questions are included in the in-class worksheets:

Knowledge-Based (Conceptual) Questions

These questions prompt students to understand the concept while the lecture is being delivered. If a student misses the explanation, the student will not be able to answer this type of question because he or she will be disoriented or disconnected from the content and eventually will lose the connection. I have found that students will try to stay focused and pay attention to the lectures to avoid such a situation. Question #1 in Figure 2 is an example of a knowledge-based question. This question requires that students pay attention to detail about one mole of a substance. They must know that the molecular mass of a substance expressed in grams is exactly one mole of that substance.

Constant-Based Questions

Several physical constant quantities are regularly used in chemistry and physical science courses: the acceleration due to gravity for Earth (*G*), ideal gas constant (*R*), Avogadro's Number (Question #2, Figure 2), Planck's constant (*h*), etc. These constant values must be recorded correctly and repeated several times through practice to confidently apply in calculations. The in-class worksheets help students to correctly record and apply these physical constants when needed as well as to serve as a resource for future application.

Calculation-Based (Mathematical) Questions

As mentioned earlier, in-class worksheets in chemistry and physical science courses will have several calculation or mathematical questions which range from simple, one-step problems to formula-based problems to multi-step, critical thinking problems. These questions are given in an in-class worksheet and solved during lecture to demonstrate step-by-step methods to find the solution and help students to remember the calculation process and replicate when needed.

Application-Based Questions

Application-based questions allow students to connect their knowledge with real life situations and application in scientific and experimental conditions. These questions also allow for interactive classroom sessions since the subject matter discussed is closely related to our everyday life or something students have heard of but have not necessarily clearly known about the underlying scientific cause. It could be a mathematical problem or a reasoning/analysis statement. It could also be a critical thinking problem and finding a connection between knowledge and everyday life. These type of questions are designed to allow students to "think outside the box" and add to the interactive classroom environment.

Critical Thinking Questions

Thought-provoking questions demand critical thinking skills which can be practiced in a classroom using in-class worksheets. Only occasionally students ask these types of questions in a class, so it will be the instructor's part to deliberately put these questions in the in-class worksheets to ensure interaction. Most of the time, students are not even aware that they are actually critically thinking. An example of a critical thinking question is: "How are buffers capable of keeping the pH just the right amount, neither too low nor too high?"

Methodology

The main method of data collection used in this study was student survey questions. There were two groups of students targeted for the survey: In-class worksheet users and In-class worksheet non-users. Each category and the respective survey is described below.

In-Class Worksheet Users

Several survey questions were used to gather feedback from the user students about their experience with in-class worksheets. Prompt and unbiased answers were expected, and students were assured that survey responses would not impact final course grades. Students were encouraged to give productive feedback so that the result will help improve future instruction and increase the quality of worksheet content (and eventually chapter content). Figure 3 gives a portion of the actual survey with first 4 questions just as the students would receive.

Non-Users of In-Class Worksheets

The scope of this research was widened to include students from another institution where in-class worksheets are not used in teaching chemistry classes. During the survey, 59 students were provided with one sample in-class worksheet used in a General Chemistry I course and explained how it is used during lecture. They were given enough time to finish reading and understanding the content of the worksheet before they were asked the survey question. While selecting the interviewees and the sample sheet, it was confirmed that the students had already finished that topic without using the in-class worksheets. This way, the students knew that the use of in-class worksheets was an alternative approach to what they had been already exposed. After thoroughly explaining the use and purpose of the in-class worksheets, one question was asked: "Do you believe that these in-class worksheets would be beneficial within your chemistry and physical science courses?"

In-Class Worksheet
Evaluation Survey

1. What physical science course are you currently taking?
2. Do you fully complete the in-class worksheet while you are in the lecture class?
3. If the in-class worksheet was not provided, how might your grade have been affected?
4. How much have the worksheets helped you to stay focused in the class?
5. How much have the in-class worksheets helped you to complete assignments?
6. How much have the in-class worksheets helped you to prepare for the exams?
7. In which area(s) have these worksheets helped you in class? (Can choose multiple answers)
8. How likely would you recommend other instructors to follow the similar techniques?
9. How likely would you suggest for this course to keep using the in-class worksheets?
10. What suggestions would you give for improvements of the in-class worksheet? (Can choose multiple answers)
11. What other improvements do you suggest?
12. What you like the most about the in-class worksheets?

Figure 3. Example questions in the student survey. Some questions include a set of choices and some are open-ended.

Student Survey Responses

The survey responses from all the participating students were collected and the number of responses in each category and the sub-category were weighted against the total number of participants regardless of the course they were taking. The percentage for each response was calculated as a percentage of total number of participants. The percentage for each semester was represented graphically for each category. Figure 4 shows the participants who completed the worksheets as directed (and who did not!) during the class time. It is interesting to see that even though the effort was made to make students complete their work, there were still some students who were not following the directions fully.

Figure 5 shows the response from students about staying focused in class from the use of the worksheets. It is good to know that even the students who responded that they do not fill out the worksheet in class felt that it would have helped them to some extent to stay focused had they used the worksheet as a medium to interact in class. This is evident when we compare the percentage for 'Mostly No' and 'Not at all' categories for Spring and Fall 2017 semesters in Figure 4 with the responses for the same semesters in Figure 5. There are several factors ranging from the student's personal, health, and family issues as well as the classroom environment and instruction methods which could influence a student's focus in class. But the objective in this study was to see the influence of the in-class work-sheet alone.

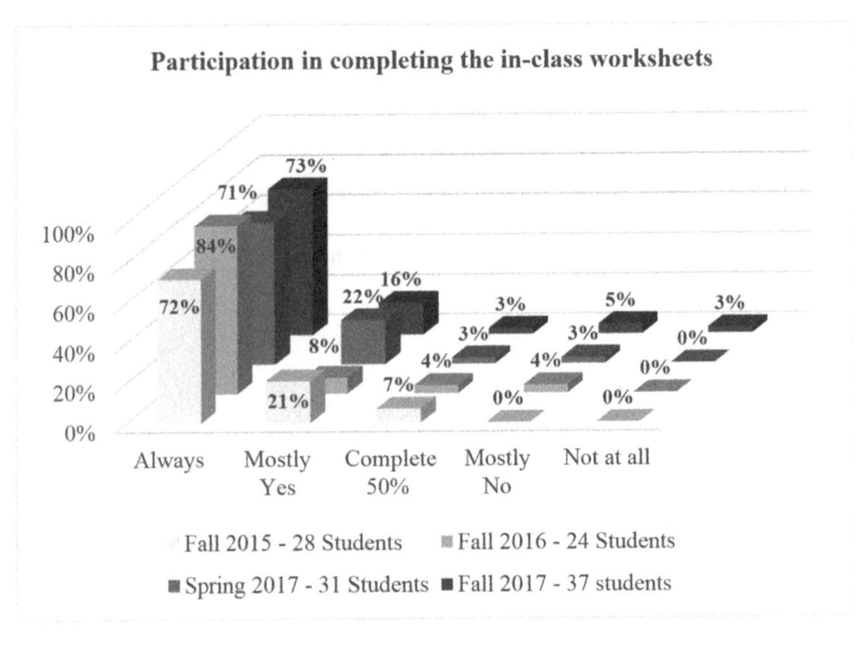

Figure 4. Percentage of participants who completed the in-class worksheets.

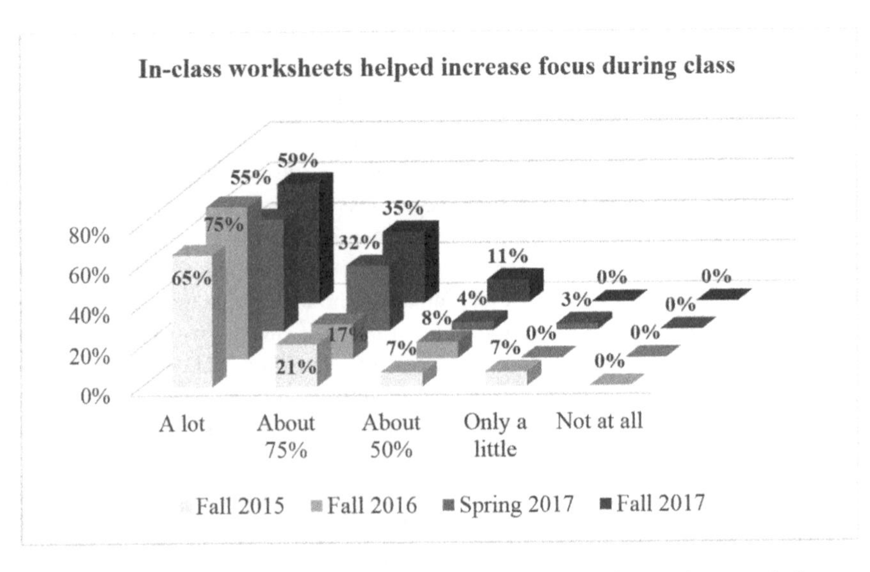

Figure 5. Responses to the question "How much have the in-class worksheets helped you to stay focused in the class?"

Assignments in these courses are given mostly after the content is completed and sometimes after finishing about half of the chapter. In any case, students get the opportunity to review the in-class worksheets and to answer the assignment questions. The responses to how the in-class worksheets would help them to complete their assignments are presented in Figure 6. A majority of students find that the in-class worksheets were very helpful for assignments. Provided the assignment questions are not typically picked up from the in-class worksheets, it is encouraging that students still find the worksheets helpful in completing their assignments.

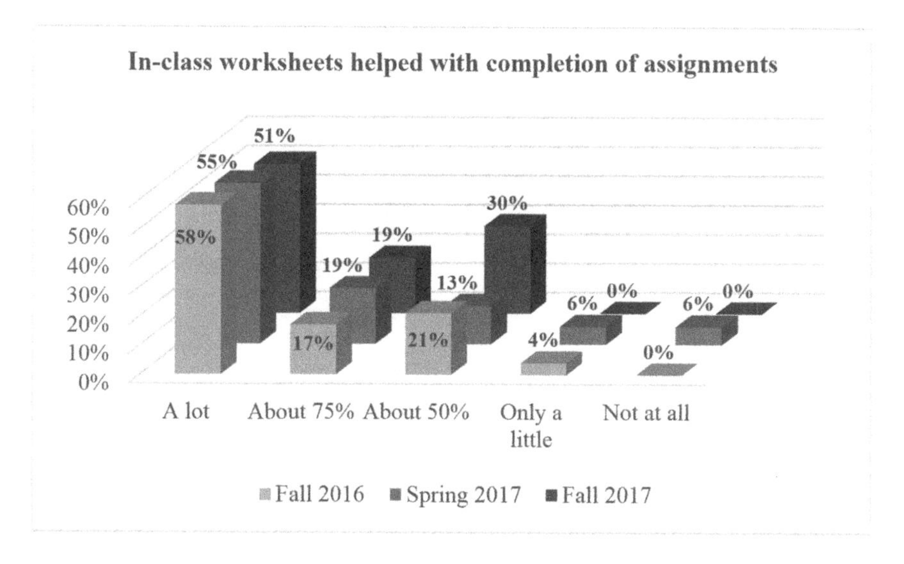

Figure 6. Correlation of concepts learned and methods used on in-class worksheets with the assignment completion.

Similar results were found for the exam preparation (Figure 7). A majority of students found that the in-class worksheets were helpful in preparing for the exams. In some semesters (Spring and Fall of 2017, for example), the percentage of students who stated the in-class worksheets were 50% or more helpful for exam preparation was above 90%. It should be noted here that the exam questions do not directly come from the in-class worksheets. However, sometimes they are picked from the assignments.

To gauge student perception of the usefulness of in-class worksheets, a set of choices were given and students could select multiple answers. Most students did not think that they could master the concept completely at the time when it was introduced (Figure 8). Mastering the concept within the class period is not generally expected either since it takes time to process the concept over a series of practice and study sessions to turn it into learned knowledge. At the same time, students thought that the worksheets were good help in solving calculation problems, to understand the concept, to solve the mathematical problems step-by-step, and to review and study for the exams. This means that on average, 70% or more of students believed that the in-class worksheets were fulfilling the purpose for which they were developed.

Student satisfaction with the in-class worksheets is reflected in the response to whether they would recommend continuation of the use of these sheets for the course. The results are very encouraging (Figure 9) since over 80% of students recommended continuing to use the worksheets. Students were so satisfied in Fall 2016 that 96% said they wanted the course to keep using the in-class worksheets.

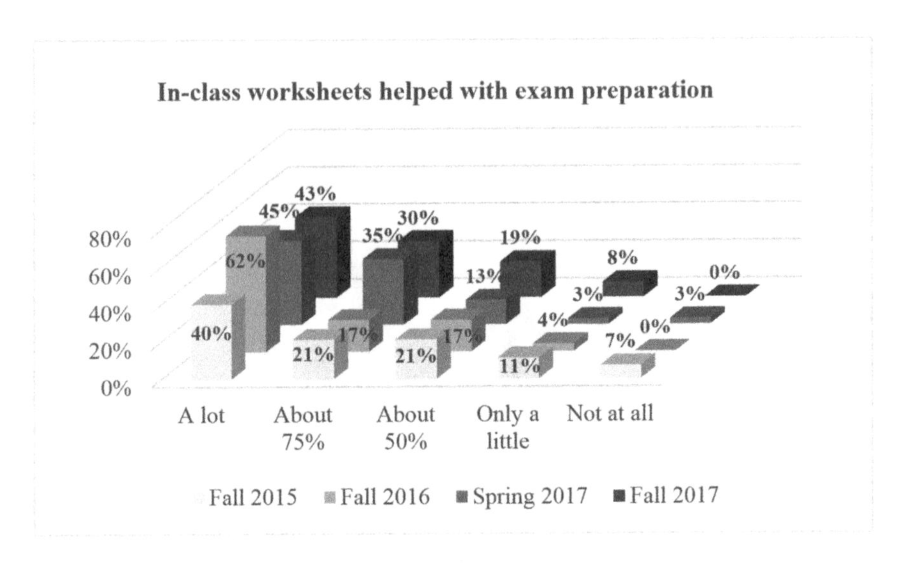

Figure 7. Usefulness of in-class worksheets for exam preparation.

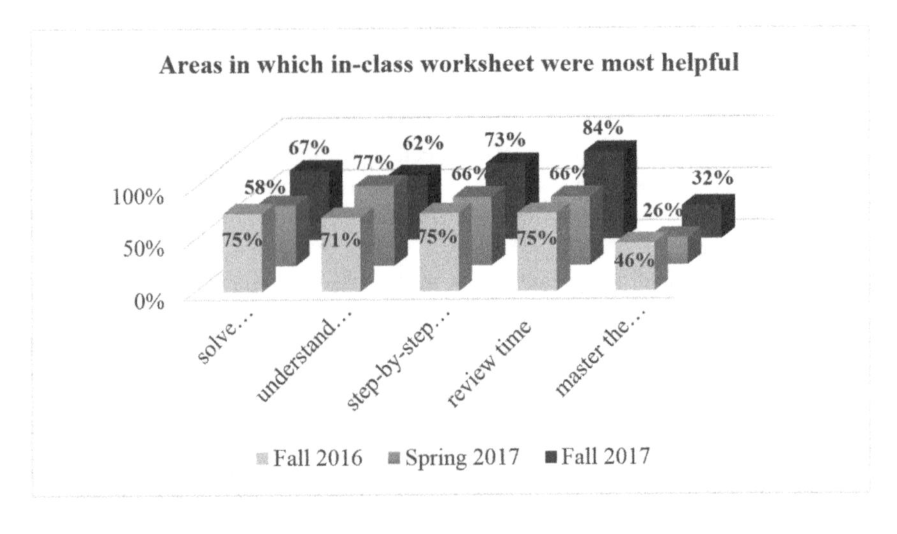

Figure 8. Areas in which in-class worksheets were useful for students.

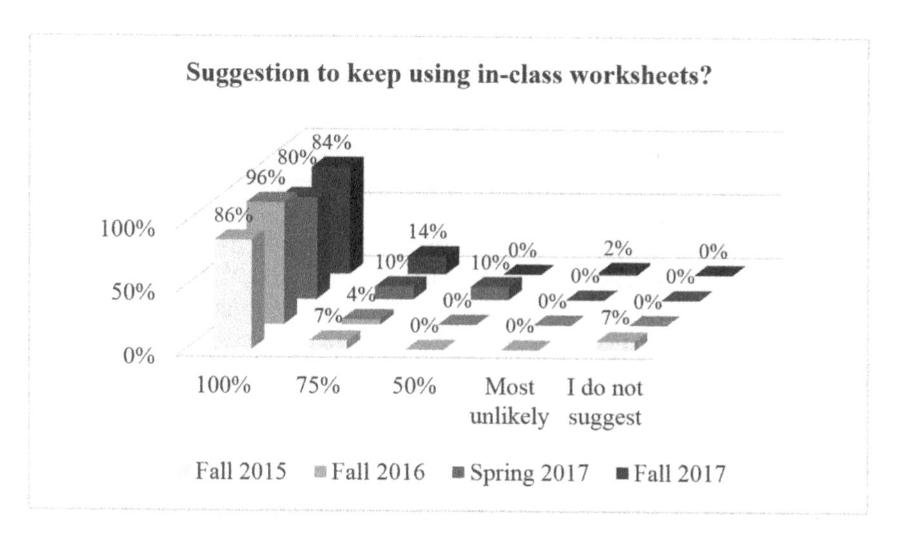

Figure 9. Positive impact of in-class worksheets validated by these student responses for continuous use in the course.

The idea of extension and expansion of helpful techniques opens up inter- or intradepartmental faculty collaboration. Since the students taking physical science courses take other courses within the department or outside, this would be a method to receive feedback from the first-hand users, the students. Do students recommend to use in-class worksheets in other courses? Responses are clearly indicating this possibility (Figure 10). With careful selection of suitable question sets to meet the student needs in that course, the possibility is unlimited.

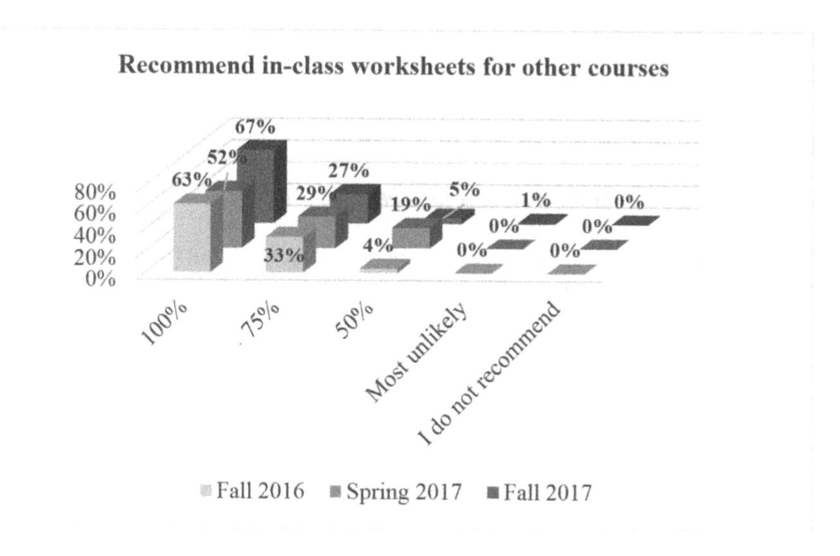

Figure 10. Responses from students about use of in-class worksheets in other courses.

The final section of the survey was to ask students for suggestions for improvement. This is my favorite section in which I get to see a visualization from the eye of a student who has spent a whole semester (sometimes more than one semester) with me in my class. There were some directed answer choices and there were open ended questions for student suggestions for improvements in the teaching process and in particular, the in-class worksheets. Directed questions gave us the responses recorded in Figure 11. While most of them did not want the the number of questions added, the same group of students then wanted more practice-based questions. This is really exciting and encouraging because the students are truly wanting more practice to be prepared for the assignments and the exams. Minor modification for the worksheets such as giving more space and making questions clearer are also suggested and are considered accordingly.

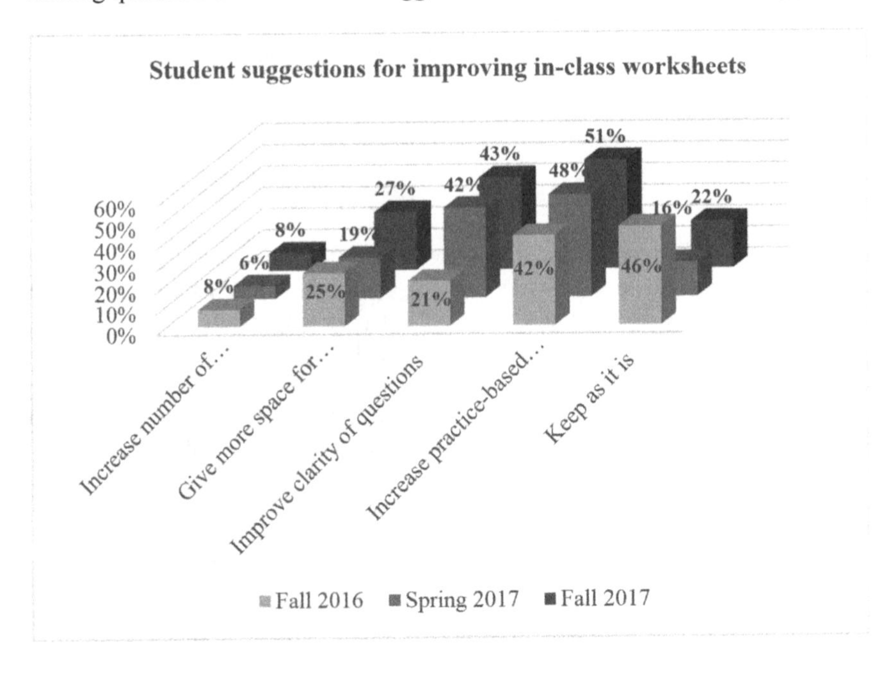

Figure 11. Student suggestions on what improvements are needed.

Most of the open ended student responses were repeated from above, but there were some truly inspiring and heart touching comments received. Some even went beyond the in-class worksheet evaluation and evaluated the instructor which was useful too. At the end of the day, consistently striving to improve and better serve students should be the sole purpose of an instructor whether it is inside a classroom or outside. Here are some example student responses:

> *"Have practice questions and in-class questions reflect each other more..."*

> *"Include more info that needs to be written down for notes (like fill in the blanks)..."*

"...not only in-class, but how our instructor encouraged participation from our class..."

"I believe they were the reason I passed my tests, studying them made a big difference..."

"Being able to refer back to for clarity as this course moves quickly..."

"Good study material..."

"help to do [homework] and assignments..."

Final Course Grades and Student Success

The grades presented here are the true grades but are presented as the percentage of all the surveyed student cohort from all physical science courses that semseter, not just one course. While grades represent the final evaluation of the student's success, the final grades under any circumstances do not represent a single factor that determined that grade. Several factors contribute to the final grades in a particular semester, such as: number of students with a good/poor background knowledge, number of class cancellations and disturbances, attendance and sickness record, financial situation of students, number of working (job) hours the students have on average, individual study habits, etc. The objective here is to focus on the use of in-class worksheets and if there was any influence/contribution on the overall grades. In some cases, the in-class worksheet might not have direct impact on the final grades since the student was already determined to work hard and make wise use of time toward a good grade. This was observed through the several survey responses. The response from a Fall 2017 student may be relevant to state here:

"I learned to be organized because you gave the worksheets and asked to keep them systematically along with the graded assignments and exams. I have a thick binder for my course now for reference."

Figure 12 shows the percentage of students who received grades A, B, C, D, or F in that semester. The number on the top of each column represents the precentage of students out of total number of students. For example, 29 on top of the very first column in the chart means 29% of students in physical science courses in Fall 2017 semester received A grade. Here, students from all the physical science courses (Physics I, General Chemistry I and II, Principles of Chemistry for non-science majors, and Physical Science) are included. Even though the difficulty level of the courses is different and the amount of mathematics involved is variable in these courses, the way of teaching (mainly due to the fact that the courses were taught by the same instructor) and the way the mathematics is connected in these courses is similar and all courses are considered among students to be the 'difficult' science courses.

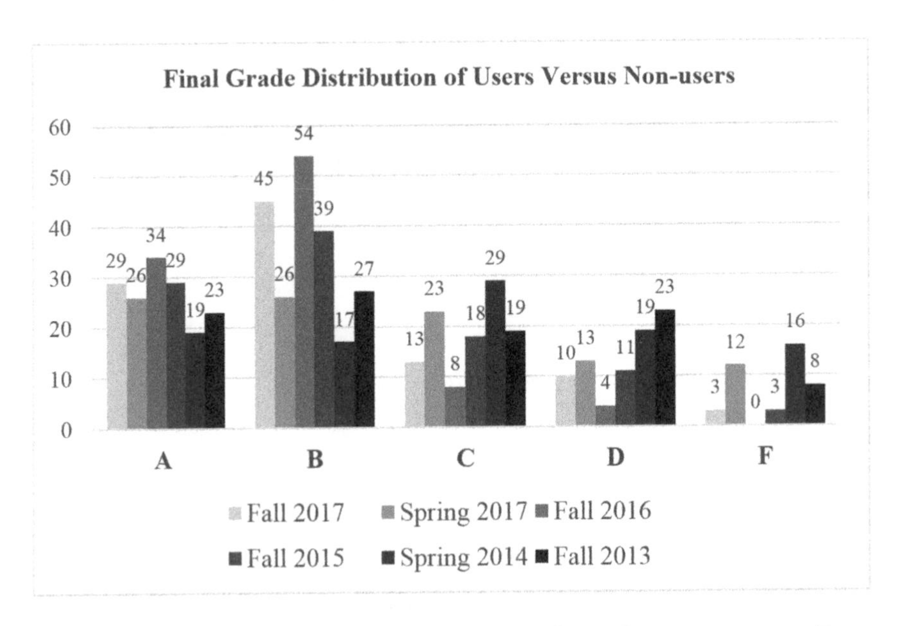

Figure 12. Final grades earned by students in physical science courses. For each level of grade, the four bars on the left represent students who used in-class worksheets during lecture, and the two bars on the right represent students who did not use in-class worksheets during lecture. The numbers represent the percentage of total students in that semester.

The first four semesters in the chart (Fall 2017, Spring 2017, Fall 2016 and Fall 2015) had used in-class worksheets for the lecture and the last two semesters (Spring 2014 and Fall 2013) did not use them.

The percentage of students who received a grade of A and B is significantly higher in the semesters where the in-class worksheets were used as compared to the semseters where they were not used: **A grade**: 19% and 23% (non-users) versus 29%, 26%, 34%, and 29% (users); **B grade**: 17% and 27% (non-users) versus 45%, 26%, 54% and 39% (users). Similarly, the percentage of students who received D and F grades in the users cohort is significantly lower as compared to to the non-users: **D grade**: 19% and 23% (non-users) versus 10%, 13%, 4%, and 11% (users); F grade: 16% and 8% (non-users) versus 3%, 12%, 0%, and 3% (users). Figure 13 shows the average percentage grade distribution between two cohorts, users and non-users. The shift towards the higher grades indicates that there is a positve impact of in-class worksheets in physical science classes.

One could argue that the surveyed students are not the same for these data from semester to semester, and in a particular semester there may be a large percentage of 'good' students than another and that could affect the percentage of the final grades. Also, it could be argued that the number of non-user group of students (Fall 2013 and Spring 2014) is comparatively smaller than the user group of students (latter four semesters). That is why the numbers used in the data are the percentages of the total number of students in physical sciences in that semester. With the same location, similar resources available for students, and similar socio-economic status, it is imperative that those factors will have minimum impact if a large group of students is surveyed over a longer period (larger number of students). Since this is not a simple comparision between two semseters, the overall progress over some years will give more accurate and dependable results. Also, in an upcoming semesters, all students in physical sciences will be tested without providing the in-class worksheets and the resulting grades will be compared.

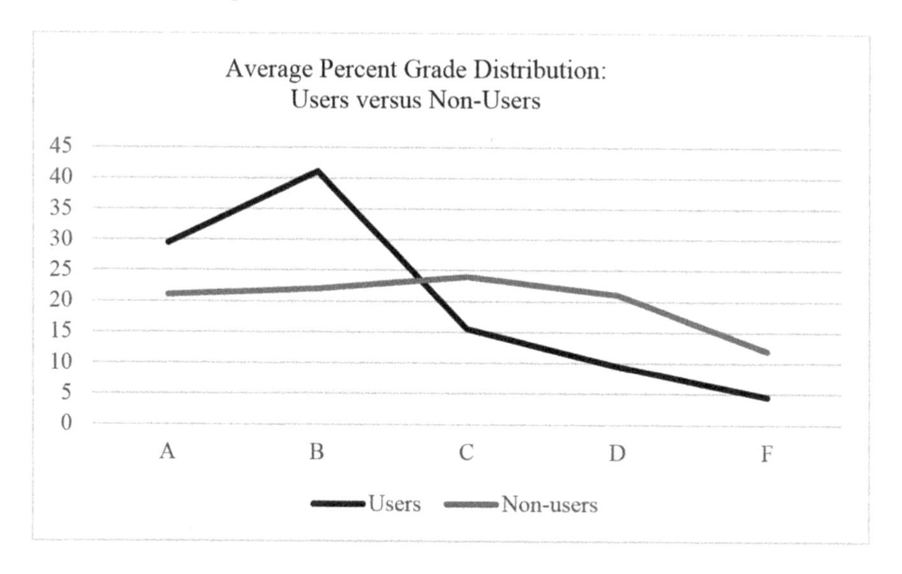

Figure 13. Average percentage grade distribution for the user group versus the non-user group.

Non-User Survey Responses

Sample responses from the non-user students when they were asked whether in-class worksheets would be beneficial also provide useful feedback:

> *"No, I don't ever take notes so these sheets wouldn't help me out. However, I am sure it would do a good job in helping those that did take notes."*

> *"Yes, I think that having these types of activity sheets during lectures would help in the class. It would help us stay focused and ensures that other students are listening to the information given by the instructor"*

> *"Yes, because how do you eat an elephant? You eat it a piece at a time! And that is what these worksheets are doing. They are managing to break down a subject that may be confusing and have long process in order to find an answer. I will be simply able to look back at my activity sheets and review the information once more to get an idea of what to do."*

"Of course this would help! I have ADHD and I know of other students who have similar situations. You could even say I have a 'straying mind' because it's difficult for me to pay attention to what the teacher says. So, these in-class calculations sheets would allow me to follow along while constantly doing something and writing notes."

"Yeah, using an activity sheet to follow along with during class is a great idea! It would reinforce our learning and keep on task. I have a tendency of losing attention halfway through the lecture and perhaps these sheets would help me in that manner."

The ratio of "Yes" versus "No" answers was a staggering 57:2 out of all 59 responses recorded from the non-user students.

Conclusion

The survey results reflect student perceptions of the benefit of in-class worksheets on their learning. Moreover, the statements made by students are the true feelings they wanted to share with their instructor and with their classmates. And the results are inspiring, encouraging the author to do something better in the future for these courses to continue to address the student needs. It becomes evident that the in-class worksheets are simple tools which are easily turned into good reference resources for students throughout the course and beyond. In-class worksheets help many students remain focused, be more disciplined and organized in their study, reduce anxiety about math-intensive courses, support understanding of calculation problems and increase confidence in their problem solving skills. This has made the class more engaged and has kept the students on the same page throughout the lecture. Final course grades suggest that the use of in-class worksheets has helped students to keep their grades up and ultimately be successful.

A teaching class is also a learning class, even for an instructor. We believe that continuous improvement and modification of in-class worksheets makes it possible to decrease the number of non-participants in class and increase the number of more organized, engaged and better-prepared students in chemistry and physical science courses.

References

1. Ozmen, H.; Yildirim, N. Effect of worksheets on student's success: Acids and bases sample. *Journal of Turkish Science Education* **2005**, *2*, 10–13.
2. Steele, K. M.; Brunhaver, S. R.; Sheppard, D. S. Feedback from in-class worksheets and discussion improves performance on the statics concept survey. *International Journal of Engineering Education* **2014**, *30*, 992–999.
3. Ambrose, S. A.; Bridges, M. W.; DiPietro, M.; Lovett, M. C.; Norman, M. K. *How Learning Works: Seven Research-based Principles for Smart Teaching*; Jossey- Bass: San Francisco, CA, 2010; Vol. 1, pp 121−152

4. Muller, P. A.; Oppenheimer, D. M. The pen is mightier than the key-board: advantages of longhand over laptop note taking. *Psychological Science* **2014**, *25*, 1159–1168.
5. Stockwell, B. R.; Stockwell, M. S.; Elise, J. Group problem solving in class improves undergraduate learning. *J. ACS Cent. Sci.* **2017**, *3*, 614–620.
6. Knight, J. K.; Wood, W. B. Teaching more by lecturing less. *Cell Biology Education* **2005**, *4*, 298–310.
7. Hake, R. R. Interactive-engagement versus traditional methods: A six-thousand-student survey of mechanics test data for introductory physics courses. *Am. J. Phys.* **1998**, *66*, 64–74.
8. Angelo, T. A.; Cross, K. P. *Classroom Assessment Techniques: A Handbook for College Teachers*; Jossey-Bass: San Francisco, CA, 1993; Vol. 1, pp 119–158.

Chapter 8

A Tool Box Approach for Student Success in Chemistry

Janice Alexander*

Department of Chemistry and Forensic Science, Flathead Valley Community College, 777 Grandview Drive, Kalispell, Montana 59901, United States
***E-mail: jalexand@fvcc.edu.**

An array of tools is needed to support the diverse community college student population. Flathead Valley Community College (FVCC) already has numerous strategies to support students in course success, degree completion and transfer including; student use of instrumentation, guided inquiry in lecture and lab, incorporating laboratory experiments relevant to multiple disciplines, grading based upon student achievement and student mentorship. More recently FVCC has seen a steep increase in the number of students entering STEM disciplines with no prior science background. Additional strategies have been added to assist this new type of student. Started as a pilot project to address a 40% fail rate in the Introduction to Chemistry course, a new course, Explorations in Chemistry was developed to bridge the gap in science background. Other engagement tactics include undergraduate research, a drop-in science tutoring center, and writing assistance for student laboratory reports.

Student Success

Research has shown large university lectures were not the initial classroom style in the United States (*1*), rather they were a byproduct of the World War II era shortage of professors compounded by scarce resources paired with increasing student numbers. As such, college faculty were not supportive of this style of learning. Over time numerous research studies compared various methods of teaching. Dubin and Taveggia showed that no significant differences emerged in content knowledge of students on final examinations based upon teaching style, and they challenged future researchers to shift the focus towards the link between teaching and learning, along with a focus on skill development. Literature of the past several decades (*2–4*) has focused on the links between teaching and learning, along with desired long term professional skills in students. With increased focus on student success and completion, along with a growing number of working and first generation college students, Flathead Valley Community College (FVCC) chemistry department has worked to implement a set of diverse strategies aimed towards improving student success and degree completion.

Background

Situated between Flathead Lake and Glacier National Park, FVCC serves a rural community of over 110,000 people distributed over 5.6 million acres in Northwestern Montana. With a headcount of 2200, 43% of FVCC students are full time and 39% of students are 25 years or older. Voluntary Framework of Accountability (VFA) data for fall 2014 first time FVCC entering students showed 59.9% of students were either still enrolled at FVCC, had transferred, or had completed their program at the end of two years (*5*). For credential seeking students this value rose to 77.4%. The chemistry department serves about 520 students a year amongst twelve different courses with 64% of students registered in either the Explorations in Chemistry or Introduction to Chemistry courses.

The chemistry department offers courses to serve both technical and transfer degree programs including:

- Brewing Science and Brewery Operations AAS
- Criminal justice AAS and transfer
- Medical Laboratory Technology AAS
- Chemistry, Biology, Engineering, Geology, and Nutrition transfer programs
- Forensic Biology and Chemistry transfer
- Pre-nursing transfer
- Pre-professional programs such as pharmacy and medicine

Course offerings include specialized courses such as Brewing Chemistry and Forensic Science, as well as traditional course sequences including the General, Organic, and Biochemistry (GOB) sequence and science major's General Chemistry sequence. In addition, the department offers an Explorations in Chemistry course serving both as a liberal arts non-majors course and as a bridge course for those without background in chemistry.. For those who need it, the first course in the GOB sequence, Introduction to General Chemistry, also serves as a stepping stone from the bridge course to the first chemistry major's course.

Across a twenty five year span many ideas have been implemented in the FVCC chemistry department. Some plans were sustainable, others have come and gone dependent upon faculty, current trends, technology, funding, college wide initiatives, and the national landscape. Common themes for student success maintained in the FVCC chemistry department have been:

- Student support mechanisms
- Collaborative learning
- Hands-on instrumentation in the laboratory
- Scientific writing
- Undergraduate research
- Faculty support mechanisms

Student Support Mechanisms

Engagement tools and methods support students and a variety of learning styles in the classroom. In addition, supporting community college students outside the classroom is critical (*6*). The need for assistance with the material, effective study skills, and navigating school and other competing demands can make the difference in persistence within a class, as well as degree completion. The chemistry department has focused on four key student support mechanisms; development of a science tutoring center, a new preparatory chemistry course, a chemistry placement project and distance learning courses.

Science Tutoring Center

Over the past five years the chemistry department has worked with other areas of the college to develop a drop in science tutoring center to supplement the long standing math and writing labs. The addition of the science tutoring center provides students support when faculty are not available, as well as providing a space where students can work with assistance available immediately when they have a question. Faculty have the opportunity to choose to devote several of the required weekly office hours working in the center to supplement the center staff. During its inaugural year October 2014 – May 2015, 82 chemistry students made 467 visits to the center for a total of 765 hours. The science tutoring center has maintained its popularity with chemistry students.

Efforts to provide chemistry support through SI (supplemental instruction) and similar mechanisms have not been sustainable. Although surveyed students express interest and provide preferred times, very few students attend the voluntary SI session. Typical comments of students refer to the difficulty in scheduling a time or recalling where to go for the SI session. The science tutoring center is permanently staffed, provides consistent and long hours, and the knowledge to assist a student across a wide variety of classes. The science tutoring center has proven to be more desirable to the students.

New Course Offerings

In 2012, thechemistry department developed and added a chemistry course Explorations in Chemistry to address an increasing student failure rate in the Introduction to Chemistry course. The student population in the Introduction to Chemistry course had become bimodal based upon student background. Unlike math and English, chemistry is not a required high school course. In addition, our chemistry curriculum lacked a series of levels to provide entry to students with diverse preparatory backgrounds. Although an additional course adds a semester to a degree plan, an unsuccessful student often will either drop out of school or change to a different major. The Explorations in Chemistry course provides students the opportunity to be successful in chemistry, maintain self-confidence, and continue on to complete their degree plan.

Appreciation for the Explorations in Chemistry course are expressed through student comments such as:

- *"The thing I enjoyed most in the class was actually learning and comprehending the material. I love how this class simplified tough concepts in a way we can understand it. I enjoyed it because I'm proud that I was able to learn and succeed. I had a miserable experience with chemistry in high school and doomed myself to a career not in the sciences."*
- *"This was my very first experience with chemistry & because I had a very patient and nice teacher who explained everything very thoroughly, I found myself enjoying chemistry a lot. It feels like solving a puzzle."*
- *"I find chemistry wildly fascinating. I have never taken a chemistry class before, so just scratching the surface of this science makes me truly appreciate the intellect of those who have made this subject a career. "*
- *"The thing I enjoyed most about this class is that I learned how to identify chemical compounds and can mentally break them down in my head just by reading the name. Reading ingredient labels is important to me and this has helped my understanding of ingredients immensely. My least favorite part is all of the math. Sometimes things that are so "black and white" and rigid exhaust my brain."*

Chemistry Course Placement

With the addition of Explorations in Chemistry to the curriculum, correct placement of students gained additional importance. Currently, the chemistry department is in the midst of a one year pilot using the Assessment and LEarning in Knowledge Spaces, ALEKS (*7*) online Knowledge Assessment module as the course placement mechanism. This module has been divided into three course placement levels. A student score on the Knowledge Assessment module will place a student into one of three courses: Explorations in Chemistry, Introduction to Chemistry, College Chemistry 1, or not ready for chemistry. A student scoring into 'not ready for chemistry' will need to increase math skills beyond the pre-algebra level prior to enrolling in any chemistry course. Students can place into Explorations in Chemistry through demonstration of pre-algebra competence through the math placement or chemistry placement system. Students that do not place into their desired chemistry course level have the option of working their way up in the ALEKS system to their desired course level or registering for the course they placed into. The goal of this ALEKS placement project is to provide a mechanism for motivated students to demonstrate readiness for a chemistry class regardless of prior chemistry background. This design also allows a student who chose to begin in Explorations in Chemistry to retake the placement test after completing the first chemistry course and work up readiness for College Chemistry I in the ALEKS system thereby skipping the Introduction to Chemistry preparatory course.

Distance Courses

Distance learning courses were added to the curriculum to serve the rural population. Creation of these courses led to recorded video lectures, use of a lightboard, and virtual office hours (8). The success of these technologies carry over into use in face to face classes and student advising essentially moving all courses into the 'blended' or 'hybrid' mode. Other areas that support student success in chemistry include advising/counseling, disability services, a behavioral intervention team, and the Veteran's Center. More faculty training, along with increasing numbers of entering students with challenges, have shifted the campus to a more student centered culture. Chemistry department faculty meet regularly to discuss students that make known or appear to need additional support. As needed, faculty will contact various services on campus for advice or assistance to support a student. The role of the faculty member from a sole focus on teaching and learning through content delivery has evolved over the years with a greater expectation for fostering student self-confidence and effective study behaviors (*9, 10*).

Collaborative Learning

Meta-analysis data has shown cooperative learning to have greater benefit to individual achievement than competitive or individualistic learning (*11*). In the FVCC chemistry classroom collaborative learning has been implemented in a variety of ways including POGIL (*12*) and Guided Inquiry (*13*), problem solving, discussion, and clicker questions.

POGIL, and its forerunner Guided Inquiry, have been used since the mid 1990's. During lecture students organize into groups of three or four and as a group they work through a series of worksheet questions. In this method the worksheet provides minimal background information and data on a subject that leads the students, through scaffolding, to develop a theory and generate equations rather than the student learning the theory through reading the textbook or learning through a traditional lecture method. Guidelines required of each group include:

1. reading the provided material prior to answering a question
2. full group participation in discussion and exchange of ideas
3. full group support until each member feels they understand and have successfully written a correct answer to the question
4. checking in with the professor at designated points.

At FVCC, professors have autonomy in the classroom, yet the chemistry department works together to provide a similar experience across multiple sections of the same course. POGIL is primarily used in Introduction to Chemistry and College Chemistry I and II courses. When using this method faculty instructors have found it most beneficial to provide closure via either a mini-lecture/discussion at the end of the class period or assign student groups to write their answers on the board, explain the steps used to arrive at their answer, and answer any questions.

The term problem solving at FVCC simply implies that students work in groups to solve problems. Essentially these are typical homework problems including conceptual, calculations, and integrated challenge questions. Implementation has included students working individually prior to organizing into groups to compare methods and results or working in groups from the beginning. Group results might be presented on the board, or correct answers provided by the instructor leading to a discussion of group answers and methods.

The goal of student success through engaged interaction in the class is also met through teaching via classroom discussion. This method allows students to provide much of the content rather than the professor. The professor becomes more a facilitator, guiding the class through the material and redirecting as appropriate.

Clicker questions follow a similar format to problem solving in regards to group work with either individuals or small groups registering their response via clicker. A graph of responses is then displayed. Students are asked to support their answer choice, leading to class discussion.

All of these engagement mechanisms provide active learning opportunities for students to learn and practice approaches to problem solving and connect students with each other in a supportive rather than competitive fashion. In addition, it allows the professor to catch conceptual and calculation errors and drive home expectations of complete thoughts in answering essay questions and showing all steps in solving a problem. This additional interaction in the classroom supports the ideal of a common goal of student success for both the professor and student.

How to find the most productive mix of approaches in the classroom is an interesting question, and left open to instructor preference at FVCC. Some professors are more comfortable with certain methods. In addition, with class sizes of twenty students or less, classes sometimes adopt a personality of their own preferring some methods over others. In addition, certain topics seem to lend themselves more readily to specific methods. There becomes a constant gauging throughout the semester by the faculty instructor of which methods to employ each day based on the topic at hand, classroom response, and classroom achievement.

Hands-On Instrumentation in the Laboratory

A strong component for many decades of our community college program has been hands-on use of instrumentation in the laboratory. The philosophy is to excite students, provide greater relevance and a usable industry skill set through student use of current technology in the field. In addition, the chemistry department feels it important to demystify the 'black box' and provide students the opportunity to gain self-confidence in their capability to learn how to use instruments.

Courses use data collection devices such as temperature and pH probes and UV-VIS spectrophotometers. Infrared spectroscopy (IR) is included in the GOB sequenceand NMR spectroscopy has been incorporated into the College Chemistry Sequence. In addition to these instruments, gas chromatography (GC) and GC-mass spectrometry (GCMS) are used in the organic chemistry sequence. Students in applied chemistry classes also use instrumentation in the laboratory. IR, GC, GCMS, and polymerase chain reaction (PCR) coupled with gel electrophoresis are used by students in the Forensic Science II class. Brewing Chemistry incorporates NMR and GCMS.

In first semester courses students are taught limited theory and analysis, along with operation and care of the instrument. Students in these courses are also taught to access library database searches. In the traditional organic chemistry 1 and 2 courses students learn significantly more theory and are expected to analyze spectra to determine purity and product identity. Through a partnership with a local high school, Organic Chemistry II students 'teach' students in a high school organic chemistry course to operate the IR and NMR spectrometers, explain and demonstrate the GCMS, and assist the high school students with spectral interpretation. Year after year, the Organic Chemistry II students express nervousness, often doubting their ability to assist the high school students. By the end of the day, the student's language has switched to amazement at their success with the task.

Funding for instrumentation is a significant challenge at a small community college. Instrumentation has primarily been funded through external grants, college foundation awards, and a college equipment fee budget. The department resource capacity has been built and maintained through a combination of new, used, and donated instruments. Replacement cycles for instrumentation have typically been 20-25 years.

Scientific Writing

The department has sustained the requirement of keeping and maintaining a laboratory notebook in all traditional chemistry courses (Explorations in Chemistry, GOB sequence, science major's course sequences). In addition, students in these courses submit ful written laboratory reports. The number of reports required per semester varies based upon the level of the course and the instructor, with a minimum expectation of four written reports. A common rubric is used throughout the department to assess the laboratory notebooks and laboratory reports. The importance of practicing and learning the methods and writing skills used by a scientist are bolstered through a similar requirement by the physics, anatomy and physiology, and biology departments.

Undergraduate Research

What began as independent study projects upon student request has over time turned into an undergraduate research program. Currently the FVCC science departments have research courses that are common course numbered with the Montana University System. This allows students to utilize their research credits upon transfer to four year institutions in Montana. Interested STEM faculty submit research project proposals each semester to the administration for approval. Upon approval, the 1 credit research course will be listed on the class schedule and included in the faculty members teaching load. A student must have instructor approval to register for the research course.

In the chemistry department, regardless of project, students are required to successfully complete a safety module for research prior to beginning any work in the laboratory. In addition students present their work at the FVCC student research conference held at the end of each semester. Past research projects have led to student presentations at state, regional, and national American Chemical Society meetings, as well as Montana Space Grant Consortium (MSGC) conferences. In addition, students have been inspired and been successful at pursuing REU summer research opportunities. Some students have immediately transitioned into undergraduate research at their transfer institution or begun working for a company interested in their research results.

Faculty Support Mechanisms

To maximize student success, both new and long term faculty need continuing professional development. Gone are the days where a new faculty member was handed a textbook and sent off to prepare to teach a course. At FVCC the chemistry department provides incoming faculty, whether tenure track or adjunct, a syllabus, daily schedule, textbook and support materials, copies of representative exams, and other relevant course material including answer keys, and assistance in setting up the LMS shell and online homework for each course. An experienced chemistry faculty member provides additional informal mentoring to new chemistry faculty. Formally, the college assigns each new tenure track faculty a mentor and requires completion of a new faculty professional development class. All new faculty, tenure-track and adjunct, have classroom observations scheduled during the first semester of teaching. These mechanisms provide substantial support to achieve a successful first experience for both students and faculty. A strong professional development mechanism at the college provides ongoing support for faculty to implement new strategies in the classroom and across the department.

Conclusion

Community college students need support and engagement inside and outside of the classroom. Some opportunities for support and engagement are mandated, while others are offered for those who choose to take advantage. This leaves inequity in support measures leading to various rates of student success. In the classroom the professor drives engagement through the mechanisms they are excited and comfortable with, thus providing a tool box with multiple options. This allows the professor to make the best choice for him/herself and provides students a greater diversity of experiences.

Research indicates that multiple strategies are required for student success (*6*). FVCC and the chemistry department have implemented this approach towards increasing student success. Many methods have been tried and those that have shown to be a best fit for the institution have been sustained creating a tool box of strategies. The next step is to analyze institutional data through a per student longitudinal lens. Regular review of tool box strategies is necessary as student demographics change and research outcomes continue to improve. Staying open to new ideas and recognizing moving targets sustains positive outcomes.

References

1. Dubin, R.; Raveggia, T. C. *The Teaching-Learning Paradox: A Comparative Analysis of College Teaching Methods*; Center for the Advanced Study of Educational Administration, University of Oregon, 1968.
2. Tinto, V. Classrooms as Communities. *J. High. Educ.* **1997**, *68*, 599–623.
3. *Development of Student Skills for Academic and Professional Success*, American Chemical Society. https://www.acs.org/content/acs/en/education/policies/twoyearcollege/7--development-of-student-skills-for-academic-and-professional-s.html (accessed December 30, 2017).
4. Schmidt, H. G.; Wagener, S. L.; Smeets, G. A. C. M.; Keemink, L. M.; van der Molen, H. T. On the Use and Misuse of Lectures in Higher Education. *Health Professions Education.* **2015**, *1*, 12–18.
5. *Voluntary Framework of Accountability*. http://vfabi.aacc.nche.edu:8080/asp/Main.aspx?rn=/20171230195748517 (accessed December 23, 2017).
6. Kinzie, J.; Kuh, G. *Review of Student Success Frameworks to Mobilize Higher Education*; Indiana University Center for Postsecondary Research. http://cpr.indiana.edu/pdf/Student%20Success%20Framework%20Report%20Kinzie%20and%20Kuh%20Edited%20Report%20for%20posting%202017.pdf (accessed December 29, 2017).
7. *Assessment and LEarning in Knowledge Spaces (ALEKS)*; McGraw Hill Education. https://www.aleks.com/ (accessed April 1, 2015).
8. Alexander, J.; Wenz, J. Serving Rural Northwestern Montana Through Online and Blended Chemistry Courses. In *Online Approaches to Chemical Education*; Sörensen, P. M.; Canelas, D. A., Eds.; ACS Symposium Series 1261; American Chemical Society: Washington, DC, 2017; pp 165−177.

9. Sorcinelli, M. D. Faculty Devolpment: The Challenge Going Forward. *Peer Review* **2007**, *9*, 4–8.
10. Sheety, A.; Moy, E.; Parsons, J.; Dunbar, D.; Doutt, K. C.; Faunce, E.; Myers, L. In *University Partnerships for Academic Programs and Professional Development*; Blessinger, P.; Cozza, B. Eds.; Innovations in Higher Education Teaching and Learning, Volume 7; Emerald Group Publishing Limited: England, 2016; Vol. 7, pp 221−241.
11. Johnson, D. W.; Johnson, R. T.; Smith, K. A. Cooperative Learning: Improving University Instruction by Basing Practice on Validated Theory. *Journal on Excellence in College Teaching* **2014**, *25*, 85–118.
12. Moog, R. S.; Spencer, J. N.; Straumanis, A. R. Process-Oriented Guided Inquiry Learning: POGIL and the POGIL Project. *Metropolitan Universities* **2006**, *17*, 41–52.
13. Farrell, J. J.; Moog, R. S.; Spencer, J. N. A Guided-Inquiry General Chemistry Course. *J. Chem. Educ.* **1999**, *76*, 570.

Undergraduate Research Opportunities

Chapter 9

Identifying, Recruiting, and Motivating Undergraduate Student Researchers at a Community College

Douglas J. Schauer*

Department of Math and Science, Southwestern Michigan College, 58900 Cherry Grove Road, Dowagiac, Michigan 49047, United States
***E-mail: dschauer@swmich.edu.**

In recent years, many community colleges have begun to embrace the value of undergraduate research. Initiating a research program at these institutions, however, presents a unique set of challenges. This chapter recounts the efforts made toward starting a research program at a community college. Special attention is given to successes in the areas of identifying, recruiting, and motivating undergraduate researchers in this setting. In particular, collaborating with the honors program on campus has led to the identification of strong research candidates who had positive undergraduate research experiences.

The Case for Undergraduate Research

For good reason, virtually every undergraduate chemistry course includes a laboratory component. By working in the laboratory, students are able to physically experience the principles that they have been taught, thus reinforcing lecture material. In addition, students gain experience in the manipulation of chemicals, critical thinking, and data analysis. The downside of the typical undergraduate laboratory experiment, however, is the contrived nature of most exercises. While most publishers have been actively updating classic experiments to reflect modern applications of chemical principles, there remains the inescapable fact that experiments in a manual are not experiments at all. Instead, the conditions for each experiment have been carefully worked out to ensure that the expected results are obtained—complete with tips on what to do if a part of the experiment fails. As such, those experiments found within a typical undergraduate curriculum fail to provide students with an authentic laboratory experience. In fact, a student who completes a typical undergraduate chemistry curriculum may well finish believing that all chemists spend their days following detailed procedures to reenact famous experiments done in the past.

The research environment provides a unique setting for learning; the questions are all open-ended, answers can rarely be found in a book, and the results are frequently unexpected. As such, students engaged in research gain all of the positive outcomes of participating in a laboratory course with many added benefits. These added benefits include experimental design, data interpretation, working independently, and persistence in response to failure. Overall an undergraduate research experience can be said to more accurately model the work of a scientist, thus illustrating a relevant application of course material (*1–5*).

Naturally, one might expect undergraduate research experiences to be most common at those institutions that already have the infrastructure in place for performing research, such as large universities with post-doctoral and graduate students. Smaller colleges have also recognized the benefits of undergraduate research and initiated similar programs for their students. The most recent institutions to conform to this trend are community colleges (*6*, *7*). This is particularly notable because the barriers to performing undergraduate research at these institutions are considerably higher than at their 4-year counterparts.

Approaches to Implementing Undergraduate Research

Undergraduate research can take many forms depending on the availability of resources. Space, equipment, funding, and faculty and staff availability factor heavily into the types of research projects that can be pursued. For the most part, undergraduate research experiences can be described according to their position on a spectrum ranging from the traditional research model to the curricular research model. These two models are described below.

Traditional Research Model

The traditional undergraduate research model closely resembles graduate research. Typically, an upper-level undergraduate student works one-on-one with a graduate student or a post-doctoral mentor. These mentors will guide the undergraduate student in their efforts toward completing a project of secondary focus to the research group. This type of experience familiarizes the undergraduate student with the research environment and culture while introducing them to sophisticated techniques and instrumentation. It also serves to draw a fairly stark contrast between the contrived course-based experiments to which they have been accustomed and the open-endedness of research experiments. The downside of this model is that it affords research experiences to a very limited number of undergraduate students due to space and time constraints. Furthermore, this model suffers from its dependence on graduate and post-doctoral students. Clearly, this makes the format more difficult to adopt at primarily undergraduate institutions (PUI), where the faculty are the only mentors available, and essentially impossible at a community colleges where faculty course teaching loads are usually heavier.

Curricular Research Model

For those institutions with a more modest research infrastructure, a curriculum-based approach to undergraduate research can be very attractive. In this approach, the number of students impacted is maximized by embedding research within the undergraduate curriculum; either a solitary experiment or a series of experiments within a laboratory course are designed to model a research project (usually inquiry- and/or project-based). This approach presents several challenges. First, by including the research in the curriculum, all students are incorporated into the project regardless of their interests or abilities. A further disadvantage of this strategy is that students will have limited one-on-one opportunities to meet with their professor to discuss the project—particularly if the entire class is keen on doing so. Also, it is considerably difficult to keep track of such a large number of projects. For this reason, curricular research usually involves the entire class working on similar projects (if not all on the same project) which diminishes the degree to which students can take ownership of their project.

Barriers to Undergraduate Research at Community Colleges

Like PUIs, two-year institutions must balance the value of undergraduate research experience with limited resources dedicated to such endeavors (*8*, *9*). Figure 1 highlights the challenges faced by community colleges beyond those at four-year universities and PUIs:

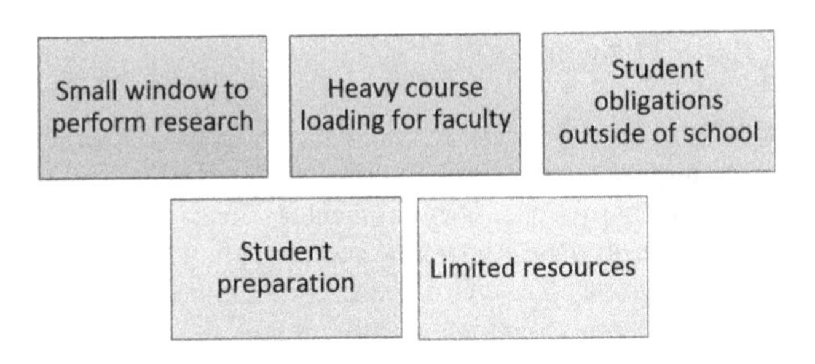

Figure 1. Factors inhibiting research experiences at community colleges.

Small Window To Perform Research

Curricula are designed to keep students at the institution for no more than 2 years, taking mostly introductory coursework. As such, there is little opportunity to evaluate students for research readiness while allowing sufficient time for significant progress to be made on a project. In the best-case-scenario, a strong candidate for research would be recognized within their first semester of a chemistry course and would have (at most), three semesters to work on their project and present their findings. The reality of the situation is that good candidates are not always recognized in their first semester. If a student does not start taking chemistry until their second year, they may only have one semester to perform research. Overcoming this barrier requires that research students be identified as soon as possible, preferably upon admission. The challenge is finding a way to identify good candidates without having met them or seeing them work in a lab.

Heavy Course Loading for Faculty

Faculty course loading is often heavy at community colleges, especially when compared to faculty counterparts at four-year universities and PUIs. This is in large part due to the lack of a research culture on most community college campuses—an artifact of the early model of community colleges where research was assumed to not be part of the faculty duties. This is a particularly daunting barrier when one considers that the faculty are the only people available to engage students in research. Furthermore, because students at a community college must be recruited earlier in their academic careers, close supervision is all the more important. While this barrier is more of a policy issue (and thus beyond the scope of this work), one can imagine that somehow integrating research with students into the current course loading policy may attenuate the demands on a faculty member.

Student Obligations Outside of School

Community colleges have a disproportionately large number of non-traditional students that may have limited time for research due to their off-campus obligations (i.e., full-time job, children, etc.). Ironically , these students also tend to be the best candidates for research due to their time management skills, maturity, and dedication. Involving these students in research would require a minimal time commitment from them beyond the time they have set aside for their courses. Invariably, this means that the research opportunity needs to somehow be embedded in coursework.

Student Preparation

As a result of open admissions and dual enrollment, the student population at community colleges represent a wide range of college readiness. For this reason, many students who are academically prepared for a particular course may not have attained a level maturity that is necessary to persevere through a research project. As such, an undergraduate research experience embedded within a curriculum must have a mechanism to select students who are up to the task.

Limited Reources

In recent years, lower enrollments at community colleges have diminished discretionary funds that may have otherwise been available for new initiatives such as a research program. This is particularly debilitating in settings that have no research culture whatsoever. Again, overcoming this barrier is more a matter of policy. However some funding can be leveraged as classroom supplies and lab fees if the research is incorporated into a course.

These unique barriers require that community college faculty need to be creative as well as resourceful in order to carry out undergraduate research. One particular challenge is finding an effective way to identify, recruit, and motivate undergraduate researchers within a community college setting.

Developing a Chemistry Research Model for Community Colleges

Previously presented work (*10*, *11*) in a technology-based chemistry program suggested that the best way to incorporate undergraduate research was through the development of a research capstone course, the benefits of which are summarized in Figure 2. The latitude to create a new course was afforded by the institution based on the recommendation of the advisory board. The advisory board was composed of representatives from local industry and was responsible for insuring that the outcomes of the program aligned with employer needs. Because of the small size of a technology-based chemistry program, the primary faculty member served as the advisor for all chemical technology students and taught all chemical technology courses. As a result, the faculty member was able to identify research students early in their academic career, sometimes prior to starting their first class. Depending on their performance in class the faculty member could include them in research activities as appropriate until it was time to take their capstone research course. This served as an excellent recruiting tool. An added benefit was that students enrolled in the course would be motivated to persevere by the grade they would receive.

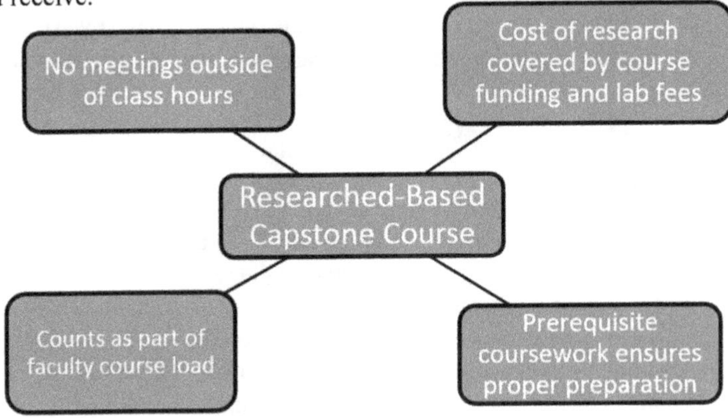

Figure 2. The benefits of instituting undergraduate research via a capstone course.

While the aforementioned model was successful within a chemical technology program, it did not translate well to a traditional chemistry program where most (if not all) of the coursework consisted of general education core courses and electives. Due to articulation agreements and federal regulations capping credit hours, the prospect of creating a research course was bleak at best. Moreover, many of the same barriers to research existed for students in the traditional program as had for the technology-based program. Clearly, a different approach was needed.

Because the student population in the traditional program was demographically very different (younger, traditional, living on-campus) from the chemical technology student population, there was hope that treating the research program as an extracurricular activity may attract research students. Efforts to recruit students in this way were initially very successful. Large numbers of students responded and were very enthusiastic at the outset. Unfortunately, this enthusiasm waned fairly quickly once the rigors of the semester took hold. Furthermore, many of the students who volunteered had not taken chemistry yet and were not academically prepared to carry out research. By the end of the semester only four students remained—none of which were recruited at the beginning of the semester. So, while introducing research via an extracurricular activity may have been an effective way of recruiting research students, it was not successful in identifying strong candidates for research, or motivating them to persevere to completion.

Traits of an Ideal Undergraduate Student Researcher

In order to more effectively identify students who have the greatest potential for undergraduate research, efforts were made to describe the attributes of the ideal candidate. Upon reflection, several traits appeared key to success:

Content Knowledge

In order to undertake research, an undergraduate student needs to have some baseline knowledge concerning the phenomenon to be studied. At the very least, some formal exposure to general chemistry (if not a full year sequence) is desirable.

Commitment

The nature of research requires that the student be prepared to set aside a significant amount of time to work in the lab. Furthermore, the student should also be flexible in that time commitment, knowing that some experiments may keep them in the lab longer than expected.

Responsibility

Because of the independent nature of research, a student accepts more responsibility for working safely, documenting procedures, managing data, and reporting findings.

Enthusiasm

The benefits of a research experience are more fully realized when a student is engaged in the project. Under these conditions, the student will be more proactive in problem-solving (thus employing critical thinking) and more likely to persevere through failed experiments.

Distant Graduation

A significant portion of the undergraduate research experience involves overcoming the learning curve corresponding to working in a new lab and on a new project. Just the process of receiving training (safety, instruments, etc.) can consume the first few weeks of lab work. Many students will not reach their "stride" until almost a semester into their project. For this reason, it is important that a student begin their research experience as soon as the circumstances will allow.

The five attributes mentioned above can be in conflict with each other. For example, a student strengthens their content knowledge by completing classes. However, each course brings the student one semester closer to graduation. With only two years being spent at a community college, each semester becomes much more valuable. Clearly, a compromise needs to be struck that maximizes content knowledge while also maximizing the length of the research experience. Often, there is also a conflict between responsibility and commitment. Some of the most responsible students are involved in extracurricular activities while maintaining jobs and caring for families. As such, these students either lack the time and/or the flexibility to carry out research. Ideally, there would be a way to identify the more responsible students prior to their taking on other commitments.

Honors Program Collaboration

In recent years, many community colleges have adopted honors programs in order to attract stronger students and reward those who take honors level classes. Many honors programs are academic units with strict requirements for membership, dedicated honors advisors, and a requirement for advanced coursework. It quickly became apparent that a collaboration with the campus honors program may serve to alleviate many of the challenges associated with identifying, recruiting, and motivating undergraduate researchers. Figure 3 outlines how collaborating with the honors program has generated a pool of undergraduate students who embody the traits of a good researcher.

While all current students are invited to join the honors program, most students begin the application process when they are still in high school. This process involves academic evaluation, letters of recommendation, and an essay. As a result, choosing researchers from the honors program ensures that the pool of students are identified early in their academic careers, have a strong academic background, and likely will have completed high school chemistry (in some cases honors or AP chemistry) prior to starting research.

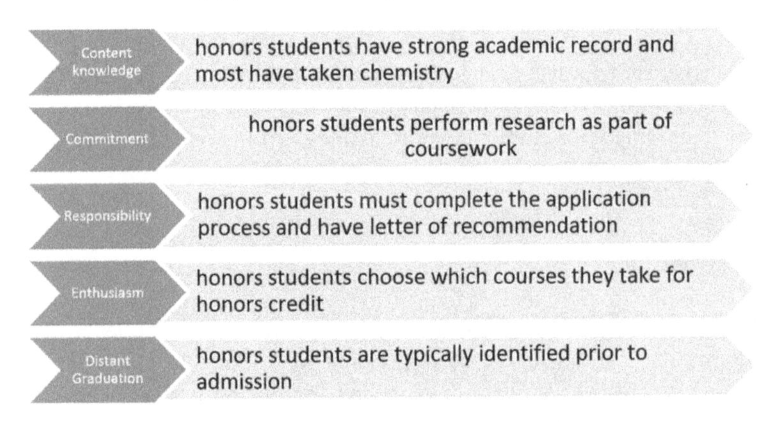

Figure 3. Traits of an ideal undergraduduate student researcher, as vetted by an honors program.

Once the students are admitted to the honors program, they meet with an honors advisor who is well-versed in the research opportunities on campus. The students are then connected to honors courses that fit their interests and majors. Students who choose honors chemistry courses are aware that undergraduate research is a major component in the honors requirement for that course. With the research embedded in their course objectives, the students can draw motivation from the fact that their course grade is tied to their laboratory research efforts. It should also be noted that students taking Honors General Chemistry I are required to take Honors General Chemistry II as part of Fall/Spring sequence.

The purpose for this is to minimize the amount of work done by the faculty in preparing research projects.

Academic performance alone is not a sole indicator for potential success in undergraduate student research. While the aim of the present work is to identify research candidates based on this parameter, it did not preclude other students from performing research outside of the honors program. In fact, some students joined the honors program after performing research while others did not have any affiliation with the honors program at all.

Honors Chemistry Curriculum

Students taking an honors chemistry class are expected to meet all of the course objectives required for non-honors students. They are also required to perform all of the same laboratory exercises. In order to receive honors credit, these students must schedule additional time to work in the laboratory on an independent research project. Typically, the students are scheduled for 2-3 hours of research every week. Each course requires that the students submit three assignments for assessment. In General Chemistry I, honors students must submit a literature review, an abstract, and present a poster at the conclusion of the semester. In General Chemistry II, the students submit a research progress report and give two oral presentations (one at an ACS conference and one at an on-campus event). Overall, the research component and these assignments count towards 20% of the student's overall grade in the chemistry course.

Upon collaborating with the College honors program, several strong students have been identified, recruited and motivated to engage in undergraduate research in several disciplines. Many of these students have had the opportunity to present their research projects at an ACS national meeting and some have gone on to successfully transfer to four-year institutions. A partnership with the honors program continues to be fostered through the development of on-campus presentation opportunities in addition to sharing the research model described herein with other disciplines across the college. The partnership has also been warmly received by administration as it has provided students with new research opportunities across the college while minimizing stress on faculty and avoiding the addition of new courses.

References

1. Lemaster, C. B. The Value of Undergraduate Research. *Chem. Educator* **1997**, *2*, 1.
2. Kimble, G. A. The Scientific Value of Undergraduate Research Participation. *Am. Psychol.* **1987**, *42*, 267.
3. Hamilton, L.; Grant, R. W.; Wolf, M. M.; Mathews, L. G. The Value of Undergraduate Research: A Study of Agribusiness Alumni Perceptions. *NACTA J.* **2016**, *60*, 207.
4. Collins, N.; Hymon-Parker, S.; Mitstifer, D. I.; Nelson Goff, B. S. Perceptions of the Value of Undergraduate Research: A Pilot Qualitative

Study of Human Sciences Graduates. *Fam. Consumer Sci. Res. J.* **2010**, *38*, 303.

5. Karukstis, K. Communicating the Importance of Undergraduate Research to Legislators. *J. Chem. Educ.* **2005**, *82*, 1279.
6. Supporting Science Research at Community Colleges. *NSTA Reports* **2015**, *26*, 1.
7. Cejda, B. D.; Hensel, N. H. *Undergraduate Research at Community Colleges*; Council on Undergraduate Research: Washington, DC, 2009.
8. Hewlett, J. A. Research at the Community College: Barriers and Opportunities. In *The Power and Promise of Early Research*; ACS Symposium Series 1231; American Chemical Society: Washington, DC, 2016; pp 137−151.
9. Lopatto, D. *Science in Solution: The Impact of Undergraduate Research on Student Learning*; Research Corporation for Science Advancement: Tuscon, 2009; pp 85−100.
10. Schauer, D. In *Embedded Research Experiences Within the Chemical Technology Curriculum: Strategies for Engaging Non-Traditional Students in Undergraduate Research.* Presented at the 246th ACS National Meeting and Exposition, Indianapolis, IN, September 8−12, 2013.
11. Schauer, D. *Research Experience for Chemical Technology Students: Identifying and Addressing the Factors that Impact Student Success.* Presented at the 242nd ACS National Meeting and Exposition, Denver, CO, August 28–September 1, 2011.

Chapter 10

Honors Modules To Infuse Research into the Chemistry Curriculum

Alycia M. Palmer* and Laura J. Anna

Chemistry Department, Montgomery College,
51 Mannakee Street, Rockville, Maryland 20850, United States
***E-mail: Alycia.Palmer@montgomerycollege.edu.**

The Chemistry Department at Montgomery College has developed and implemented strategies to help students navigate and successfully complete the chemistry curriculum and transfer to a four-year program. Students at two-year colleges are often looking for opportunities to gain experience outside of the traditional classroom and to engage in independent research projects. The Chemistry Department has adapted an honors course module, designed to pair honors enriched coursework with a standard class section, into an opportunity to introduce students to undergraduate research. This model has been integrated into both general chemistry and organic chemistry courses. The long term goal for introducing this research-based honors module is to increase student success for degree completion with increased transfer success and to expose students to scientific research.

Introduction

Much of the research done by undergraduate students occurs at bachelors-granting institutions, government facilities, or in industry. The purpose of this research is often to give students exposure to the laboratory environment with the added benefit that students may be able to contribute toward a publication. However, research at two-year colleges has a different purpose. Due to a smaller amount of resources, limited equipment and a reduced pool of students with only a semester or two of STEM courses, the research being performed at two-year colleges tends to focus on practicing the scientific method rather than generating publication-quality data. Authentic undergraduate research experiences, where students work on a problem with an unknown solution, have been shown to sustain motivation for students to complete a STEM degree (*1*) and also retain underrepresented minorities in the field of STEM (*2*). Here we highlight how authentic undergraduate research has been infused into an undergraduate curriculum at a two-year college for general and organic chemistry students.

Motivation for Research at Two-Year Colleges

Although two-year college students are not typically involved with publication-quality research, there are many reasons why they should be involved in a research project. First, practicing the scientific method and generating results in a lab setting connects them to science in a real way. Here, students confront the erroneous belief that science requires exceptional ability. Over the course of even one semester, they learn that success in lab comes from following the scientific method, where failure translates to new information and persistence is rewarded with meaningful data. By changing students' perceptions about science requiring exceptional ability, there is potential for increasing retention in STEM courses. One study investigated how faculty sharing struggle stories with students changed how the students thought about their own ability in science (*3*). The struggle stories relate real-life scenarios that famous scientists had to overcome before they made break throughs, stories of not having enough income or exhaustively working on the same project for years until they solved their research problem. When students learn that even Einstein struggled, they are more likely to look at their own struggle as something that can be overcome. This is especially important for retention of STEM majors at community colleges, where many students are worried about finances, dealing with placement into developmental courses, and trying to manage time between work and school obligations. By recognizing the difficulties that many community college students face, we can highlight those students who have learned to cope and persist through challenging times as having strong potential for a career in science. In fact, it has been shown that one predictor of persistence in a STEM major is the ability to cope with failure and frustration (*4*). The students who participate in a research module get to become a scientist during that time, and in this role working alongside a mentor, students begin to see that their effort leads to real progress toward solving a scientific problem.

Scientific research may also be an avenue for retaining underrepresented minority students (URMs) in STEM (*5*). One study tracked 17,000 students in the US who were declared STEM majors and found that after 6 years, only 41% of those students, on average, obtained a STEM degree (*6*). Even more troublesome is that within the subset of URMs from the study, only 20% obtained a STEM degree. The inability to retain underrepresented minorities in STEM degree pathways has been refered to as the leaky pipeline (*7*). To recruit and retain URMs in STEM, many initiatives are taking place nationwide, including novel advising efforts, active learning in the classroom, and undergraduate research (*8*, *9*). By nature, research modules are a form of active learning, and by being paired with a faculty mentor, students have access to expert advice about appropriate courses in their career pathway and continuous support in times of doubt. A faculty mentor can also be available at crucial points in a student's education to decouple temporary setbacks, such as a low exam score or a failed experiment, from a perception of the student's own ability (*7*).

Studies have shown that involvement in a research project may also lead to retention of minority STEM students (*5*). Furthermore, it has been shown that more hands-on research is better (*4*), which makes programs where first year undergraduate students are able to participate in research, such as the Research Experiences to Enhance Learning (REEL) program at The Ohio State University, so valuable (*10*). Being part of a research module also benefits students by engaging them with course content outside of lecture. Revisiting the material originally presented in lecture helps students retain the information, and the extra time spent actively solving a problem related to the material helps increase their understanding (*9*).

Not only does participation in a research module help students with chemistry-related skills, but many soft skills are also developed over the semester, such as time management, communication of ideas, and goal setting. By working closely with a faculty mentor to build these skills, students are better prepared for transfer to a four-year institution.

Montgomery College, a two-year college composed of three campuses in the metropolitan Washington D.C. area, serves over 24,000 students in credit-level coursework. Our course offerings include introductory chemistry, a two-semester general chemistry course sequence, and a two-semester organic chemistry course sequence along with a one-semester general chemistry course for engineering students and general education chemistry courses for non-science majors. Class sizes range from a maximum of 24 students in general chemistry to 32 students in organic chemistry, and all courses have linked laboratory sections of 24 and 16 students respectively. The Rockville campus chemistry department is the largest of the three campuses with twelve full-time faculty members, six full-time laboratory and office staff members, and part-time faculty instructors serving 900-1000 students in chemistry courses every semester.

Montgomery College is unique with respect to student diversity. A 2015 Chronicle of Higher Education report named Montgomery College the ninth most diverse public two-year school in the nation (*11*). The Community College Review diversity index score for Montgomery College is 0.78. This score is a measure of the probability that any two students selected at random are from different ethnic groups; a diversity score closer to 1.00 indicates a more diverse student body (*12*). Our diverse student population represents over 160 countries with no majority race (27% black, 24% Hispanic, 24% white and 11% Asian).

The degree of diversity at Montgomery College presents faculty with the opportunity to recruit and retain underrepresented students in STEM programs. As a college, many initiatives are in place, such as embedded classroom support with Learning Assistants and support through the Achieving the Promise Academy to serve all students in high enrolled, high risk STEM courses. Here we discuss efforts within the chemistry department to reach students through an honors research module and independent scientific research.

Research Opportunities and Learning Outcomes

Students have two options for completing research at Montgomery College, a course-connected honors module, CHEMxxxHM, or an Introduction to Scientific Research course, SCIR 297. If students choose the honors module, the research is completed as an extension of a regularly offered course, and faculty are compensated for mentoring each honors student. For instance, honors general chemistry has the same learning objectives as the non-honors course, but it has an additional honors component. A student's eligibility requirement to participate in the honors course option is a 3.2 GPA and a minimum 12 credit hours completed at the college. However, exceptions to the minimum credit hour requirement can be made, particularly for students enrolled in the first-semester general chemistry course honors option. Student learning outcomes for participation in an honors module include:

- To be part of a community of students who are like-minded and similarly motivated
- To be part of a class with people who like to engage in ideas
- To be part of a class that will challenge your skills, beliefs and enable you to grow intellectually and academically
- To explore course material not covered in a non-honors section or to cover some topics more deeply
- To conduct research that may help define future education or career paths
- To form closer relationships with instructors as well as other students
- To provide the student with extra-curricular or co-curricular activities not associated with a non-honors course

If students choose to enroll in the Introduction to Scientific Research course, SCIR 297, the research is completed as the primary objective of the course. This course is designed for students enrolled in science, engineering or mathematics (SEM) programs to build upon skills learned from introductory SEM courses so as to generate competency in scientific critical thinking and research. The SCIR297 course enables the student to pursue a research topic of their own choosing with the guidance and supervision of their faculty mentor. This option is suggested for students who are interested in careers in science as well as students who are accepted into external research internships and wish to receive college credit for that opportunity. Montgomery College students benefit from the close proximity of a number of government and biotechnology laboratory facilities in the local area, including the National Institute of Standards and Technology (NIST) and the National Institute of Health (NIH). The student eligibility requirement for SCIR197 is a minimum 3.0 GPA with course pre-requisites of Principles of Chemistry I, Principles of Biology I and Precalculus. SCIR 297 is a 2-credit course which includes a minimum of 3 hours a week of laboratory work and 1 hour per week of a seminar-style class meeting led by a faculty member. Students may repeat SCIR297 for a maximum of 6 credits, and students may transfer up to 4 of these credits as a lower-level science elective to one of our flagship four-year institutions.

Through the learning outcomes of SCIR297, students will be able to:

- Analyze research articles in refereed journals
- Apply research guidelines regarding record keeping
- Demonstrate an understanding of the four essential elements in any subject being studied characterization, hypothesis, predictions, and experiments
- Identify a subject to study and find ways to study it

Features of the two research opportunities are summarized in Table 1.

Table 1. HM course module and SCIR297 requirements

Course Honors Module, CHEMxxxHM	***Introduction to Scientific Research, SCIR297***
Enrollment eligibility requirements	
• 3.2 GPA, 12 credit hours • Pre-requisites: same as course, instructor approval	• 3.0 GPA • Pre-requisites: Biologiy I, Principles of Chemistry I, Precalculus; instructor approval
Total student enrollment	
• 45 students since Fall 2015	• 8 students since Fall 2015
Faculty compensation *ESH = equivalent semester hour*	
• 0.2 ESH per student • 1 ESH max for faculty	• 0.5 ESH for 1-2 students • 1.0 ESH for 3-5 students
Student credit load	
• No additional course credit	• 2 credits, repeatable up to 6

The majority of our students have elected to participate in a course-connected honors module citing the advantage that this model provides an authentic laboratory research experience without additional tuition cost. The course-connected HM offers the flexibility to be combined with any course, allowing an opportunity to students that may not have otherwise considered participating in research. The SCIR297 course option provides the benefit of course credit for more in-depth research projects and for students participating in external internships and offers transferability to one of our flagship 4-year institutions, which includes the University of Maryland. Whether students choose to partake in research through a course-connected honors module or SCIR297, both pathways have the same general learning outcome: to solve a scientific problem by practicing the scientific method.

Honors Module Research Model

According to the Montgomery College Honors College, an honors module is defined as an opportunity where students take the regular non-honors course while working on an additional independent study topic or a research project with the course instructor. The Chemistry Department has developed the honors module into an authentic undergraduate research experience. The honors module research model divides the honors curriculum component into three units, (1) background research and project design, (2) laboratory research, (3) data analysis and final presentation.

During the first unit of background research and project design, the student and faculty mentor meet one-on-one to discuss potential projects, decide on final project, review background literature and design the experimental protocol for the project. Montgomery College has access to many of the major scientific journals, and the library offers training for students to learn to navigate the library journal database.

Once the experimental protocol has been developed, the student and faculty mentor will schedule time to conduct the hands-on laboratory component of the project. This laboratory time is in addition to the regular laboratory schedule associated with the course. Depending on the project, the additional lab work may range from a 2-4-hour time block once or twice a week for 3-4 weeks or longer. There is variability in this lab time commitment dependent on schedules and progress of the project. Given that students do not receive additional credit for the honors module work and that faculty compensation for mentorship is minimal, both faculty and students engaged in HM projects are encouraged to keep the laboratory component to a manageable time commitment.

The last unit is performed outside of the laboratory space and allows the student to spend time analyzing data that was collected and to organize the results to present to peers and faculty. At the end of the academic year, the Department hosts a student research poster and presentation social event. Students that have completed an HM project or the SCIR297 course present their research in the form of a short presentation, either orally to an audience or in front of a poster that they create. The Department provides food and an encouraging environment where faculty, staff and students can socialize and provide support for all student participants. Additionally, a new college-wide event initiated in Spring 2017, the Undergraduate Research Conference hosted by the College's STEM Vice President and Provost, has allowed our students to present in a larger venue and with audience members from outside of the MC community. Guest speakers from industry and neighboring research universities were invited to give a keynote address followed by a panel discussion to share advice to students transitioning into careers and four-year programs in STEM. Posters and presentations were judged and awards were presented to recognize the best student presentations.

Student Research Project Scope

At least half of the twelve full-time chemistry faculty members at our campus have active research interests to support undergraduate research. Faculty mentors typically recruit students from the chemistry course sections they are teaching. These students range from beginners in the Principles of Chemistry I course to more advanced students taking the 200-level organic chemistry course sequence. Student registration in an honors module may occur up to two weeks into the start of a term and still allow students sufficient time to complete an honors module research project. This option allows faculty to recruit their own students at an early point in their education. Faculty also recruit students at the end of the term who may be interested in pursuing an honors project in a subsequent chemistry course.

When faculty and students meet to select a research project they may be inspired by recent events in the news, a current experiment being performed in a chemistry course, or a previous project that the faculty mentor has experience with. Regardless, the research topic must meet safety requirements and undergo careful planning with the laboratory staff to ensure the project will not be harmful to the student, will not be cost prohibitive, and that the waste generated can be safely disposed of.

Access to instrumentation and laboratory resources can often be a barrier to initiating an undergraduate research program at a two-year college. The Chemistry Department at the Rockville campus is fortunate to be equipped with a dedicated laboratory space for student projects as well as an instrumentation room. Analytical instrumentation within the department includes an atomic absorption spectrophotometer, GC-MS, HPLC, FT-IR spectrometer, 90MHz NMR spectrometer, stopped flow spectrometer and an X-ray diffractometer. Full-time laboratory staff members assist with regular maintenance, set-up and trouble-shooting of the analytical instrumentation. Many of the HM and SCIR297 student projects incorporate instrumental analysis, giving students additional experience with instrumentation techniques not used in the regular chemistry laboratory curriculum.

Participation in a research module benefits students by engaging them with course content outside of lecture. Many of the research projects can be closely aligned with course topics in general and organic chemistry. Example projects related to general chemistry have included using atomic absorption to analyze food products, water quality and environmental samples. Example projects related to organic chemistry include incorporating green chemistry such as microwave-induced syntheses and include IR and NMR analysis of products. Added exposure to material presented in lecture will help students retain the information, and the extra time spent actively solving a problem related to the material may help their understanding.

Success of Students Involved in Research

Since the introduction of the honors module research model in the Fall 2015 term, forty-five students have participated in a course-connected honors module research projects. The SCIR297 course was also introduced in the Fall 2015 term and eight students have completed research projects through this alternate course offering.

Of the forty-five HM students, 73% participated in a research project connected to their general chemistry I and II courses. Additionally, 30% of these students were either in a non-traditional STEM degree program or in the General Studies degree program. The option to have a research experience linked with a first or second term general chemistry course allows this high impact opportunity to be available to students who may not have traditionally considered engaging in research.

A smaller percentage of students, 27%, enrolled in an HM section connected with their 200-level organic chemistry, and all of these students were enrolled in an AS Science degree program. Student majors enrolled in course-connected honors modules and SCIR297 are summarized in Table 2.

Table 2. Degree programs of students enrolled in course-connected honors modules and SCIR2917

Course/Student major	*Gen. Chem I*	*Gen. Chem II*	*Org. Chem I*	*Org. Chem II*	*SCIR297*
AS Science – Chem/Biochem	1	5	1	2	1
AS Science – Life Sciences	5	3	3	1	1
AS Engineering	1	4	--	2	3
AS Science – Physics	1	--	--	--	--
AS Science – Mathematics	1	2	--	1	1
Nursing	--	1	--	--	1

Continued on next page.

Table 2. (Continued). Degree programs of students enrolled in course-connected honors modules and SCIR2917

Course/Student major	*Gen. Chem I*	*Gen. Chem II*	*Org. Chem I*	*Org. Chem II*	*SCIR297*
Computer gaming/ditigal media	2	--	--	--	--
General Studies	6	1	1	1	1
Total number of students	**17**	**16**	**5**	**7**	**8**

All fifty-three students that participated in a course honors module or SCIR297 successfully completed their respective chemistry course with a grade of C or better. Figure 1 shows the distribution of final course grades of students in each chemistry course and SCIR297.

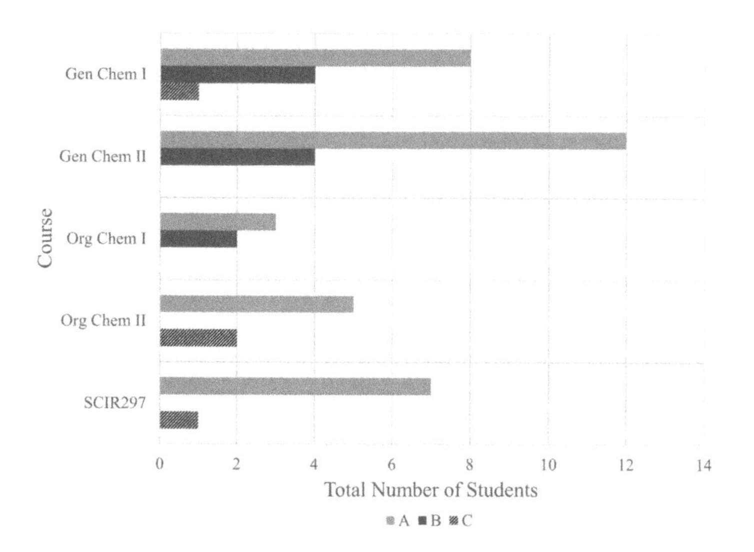

Figure 1. Final course grades of students participating in an HM project or SCIR297project.

In addition to their course success, eight students continued to work on their research projects by either continuing to participate in an honors module connected to a subsequent chemistry course, or elected to register for the SCIR297 course to explore their project in greater depth. Seven of the HM students were accepted to an industry internship. We do not have data to track student involvement in research after they transfer to four-year institutions. Such data is difficult to acquire because it requires a continuous line of communication with students after they transfer and for students to self-report their involvement.

Challenges and Sustainability of the HM Research Model

There are many challenges to introducing undergraduate research opportunities at the two-year college level. Lack of institutional support, heavy faculty teaching loads, and limited faciltity and funding resources can be barriers to successful undergraduate research programs. Constraints on students, including time and economic challenges, may prohibit them from participating in a research experience. The honors module research project design takes into account the limited time that many two-year college students and faculty have to devote to a research project, yet still offers an authentic research experience. Additionally, due to safety concerns, students performing research in a lab setting must be directly supervised by a faculty member. This requires a significant amount of time on the part of the faculty mentor. Due to differences in scheduling, it is often not possible to schedule all honors students to complete their research at the same time. Although more time is required on the part of the faculty member, it makes for a richer experience since faculty and students are performing research together. Many faculty members find this style of undergraduate research mentorship to be a rewarding experience.

With sustainability in mind, many faculty have embraced the course-connected honors module as an extension of their teaching pedagogy to promote student success. Faculty mentors consider project interests with available instrumentation and many student projects relate to and expand upon the linked course curriculum content. Additionally,several students' research projects have resulted in new or improved experimental procedures in the laboratory course curriculum.

Conclusion

A course-connected project-based honors module was designed to provide students with an authentic undergraduate research experience in all levels of chemistry at Montgomery College. Advantages of conducting research through a course honors module were that students did not need to pay any additional tuition charge for the honors unit and it offered students a research experience early in their academic career. The model also allows students that may not be a declared STEM program to participate in and experience a research opportunity. Undergraduate research programs are typically marketed to students in STEM programs with limited opportunities for students in non-STEM pathways. With general studies being the highest enrolled degree program at MC, exposing these students to a research experience may be a way to recruit a diverse population of students to consider and pursue STEM careers.

An introduction to research early in a student's education builds confidence that stays with them as they continue in their course studies. All students engaged in a research project were successful in their respective co-enrolled chemistry course, and several students went on to successfully complete multiple semesters of either HM or SCIR297 research projects or external research internships. Our course-connected honors module research model allowed students enrolled in non-STEM programs and general studies degree pathways who may not have considered or completed research at four-year college to engage in a high-impact research opportunity.

Acknowledgments

The authors would like to express much gratitude to the college-wide honors coordinator, Lucy Laufe and the MC Honors College. We also thank all department faculty, including those who mentored students with research projects as well as anyone who supported our students at presentations. We appreciate the time spent by staff members to order supplies, assist with laboratory resources and help with instrumentation. Finally, we thank the students involved in research for their enthusiasm and willingness to partake in research and to present their results to their peers.

References

1. Maton, K. I.; Hrabowski, F. A. Increasing the Number of African American PhDs in the Sciences and Engineering: A Strengths-Based Approach. *Am. Psychol.* **2004**, *59*, 547–556.
2. Hurtado, S.; Cabrera, N. L.; Lin, M. H.; Arrelano, L.; Espinosa, L. L. Diversifying Science: Underrepresented Student Experiences in Structured Research Programs. *Res. High. Ed.* **2009**, *50*, 189–214.
3. Lin-Siegler, X.; Ahn, J. N.; Chen, J.; Fang, F. A.; Luna-Lucero, M. Even Einstein Struggled: Effects of Learning About Great Scientists' Struggles on High School Students' Motivation to Learn Science. *J. Educ. Psychol.* **2016**, *108*, 314–328.
4. Harsh, J. A.; Maltese, A. V.; Tai, R. H. Undergraduate Research Experiences from a Longitudinal Perspective. *J. Coll. Sci. Teach.* **2011**, *41*, 84–91.
5. Russell, S. H.; Hancock, M. P.; McCullough, J. Benefits of Undergraduate Research Experiences. *Science* **2007**, *316*, 548–549.
6. Waldrop, M. Why We are Teaching Science Wrong, and How to Make it Right. *Nature* **2015**, *253*, 272–274. https://www.nature.com/news/why-we-are-teaching-science-wrong-and-how-to-make-it-right-1.17963 (accessed March 31, 2017).
7. Hernandez, P. R.; Schultz, P. W.; Estrada, M.; Woodcock, A.; Chance, R. C. Sustaining Optimal Motivation: A Longitudinal Analysis of Interventions to Broaden Participation of Underrepresented Students in STEM. *J. Educ. Psychol.* **2013**, *105*, 89–107.
8. Tovel-Lindsey, B.; Levis-Fitzgerald, M.; Barber, P. H.; Hasson, T. Increasing Persistence in Undergraduate Science Majors: A Model for Institutional Support of Underrepresented Students. *CBE Life Sci. Educ.* **2015**, *14*, 1–12.
9. Lopatto, D. Undergraduate Research Experiences Support Science Career Decisions and Active Learning. *CBE Life Sci. Educ.* **2007**, *6*, 297–306.
10. Clark, T. M.; Ricciardo, R.; Weaver, T. Transitioning from Expository Laboratory Experiments to Course-Based Undergraduate Research in General Chemistry. *J. Chem. Ed.* **2016**, *93*, 56–63.

11. *Colleges with the Greatest Racial and Ethnic Diversity,* Fall 2015. https://www.chronicle.com/article/Colleges-With-the-Greatest/240582 (accessed Dec. 28, 2017).
12. *Community College Review. Diversity in U.S. Community Colleges (2017-2018)*. https://www.communitycollegereview.com/diversity-rankings/national-data (accessed Dec. 28, 2017).

Chapter 11

College Students Get Excited about Whiskey: The Pseudo-Accidental Creation of a Thriving Independent Student Research Program at a Two-Year Community College

Regan Silvestri*

Division of Science and Mathematics, Lorain County Community College, 1005 North Abbe Rd., Elyria, Ohio 44035, United States
***E-mail: rsilvestri@lorainccc.edu.**

An almost chance encounter between a chemistry professor and the founder of a start-up whiskey company would quickly lead to the pseudo-accidental creation of an independent student research program in analytical chemistry for undergraduates. The entrepreneurial start-up company Cleveland Whiskey, founded on the basis of a revolutionary production technology, produced new experimental flavors of bourbon whiskey including black cherry, apple, hickory, sugar maple and honey locust. Students at Lorain County Community College used gas chromatography-mass spectroscopy to identify and profile the distinct flavor compounds in these unprecedented flavors of bourbon whiskey. In the process, a vibrant independent student research program for undergraduates has been created and is blossoming. Students are enthusiastic to become involved in the "flavor of whiskey" project which has quite the buzz on campus. The student research has since expanded to include characterization of the flavor compounds in experimental samples of Chinese baijiu, a sorghum spirit, which has been wood aged in an effort to impart flavors akin to American whiskey. Most recently, the student research has evolved to process variable optimization for the infusion of bourbon into barbecue smoker bricks. Growth of the research program has been made possible with grateful thanks to funding via grant awards from multiple sources. This is the story of how a successful student research program can be put together at a two-year community college: from inception, to involvement, to funding and sustainability.

Introduction

An almost chance encounter between a chemistry professor and the founder of a start-up whiskey company quickly led to the pseudo-accidental creation of an independent student research program in analytical chemistry for undergraduates. A research group based on the graduate school model has been established at a two-year community college, with students performing substantial project-based independent research on an ongoing basis within the structure of a research group working on cooperative topics. I would like to share the story of how all of this happened, as when we get to the end of the story there are, I believe, some valuable lessons learned.

Whiskey 101

Whiskey is made by a process where grain mash is fermented converting sugars into ethanol and then distilled to purify the fermentation products into a clear spirit. Traditionally, the clear spirit is then aged in charred oak barrels for up to 10 years or more. During this time the spirit becomes flavored and colored with compounds that leach into the spirit from the charred oak barrel. To be labeled as bourbon, the spirit must be aged in a new (virgin) charred oak barrel (*1*).

As a result of the process of maturing with wood, whiskeys are chemically complex mixtures of flavor compounds including esters, aldehydes, lactones, phenolics, and other alcohols. Essentially, whiskeys are chemically complex mixtures of a large number of flavor compounds, all present in very low quantities. Therefore, the flavor and fragrance industry recognizes gas chromatography-mass spectroscopy (GC-MS) as a routine technique for studying these complex flavor mixtures, but I'm getting ahead of the story.

An Incidental Meeting

I invest a great deal of time staying in contact with alumni. During one lunch, a former student was telling me of the fabulous experience he recently had during an internship at a whiskey company while completing his master's degree. "A whiskey company would need chemists," I thought. "Whiskey," I thought, "how cool is that?" A simple thought that would ultimately prove central to the story. I expressed that I would like to organize a tour of the whiskey company for my current students and he agreed to put me into contact with his supervisor. More than a year had passed when I eventually organized a tour of Cleveland Whiskey as a monthly meeting of the Cleveland Section of the Society for Applied Spectroscopy. You should already understand from the tone of the story that nothing here was planned more than one step (maybe two) in advance, a precedent that would persist.

I arrived early that evening since I had arranged the tour and was pleasantly surprised when the door was answered by the founder and CEO of the company himself, Tom Lix, who would be our tour guide. I introduced myself, and when I told him that I was a Professor of Chemistry at Lorain County Community College, his immediate reply was that we should cooperate on a joint research project using GC-MS to study some new whiskey flavors he was experimenting with. How did he know about GC-MS? He is not an analytical chemist. I immediately replied, "Yes, let's establish a joint research project." Honestly, I didn't know what we would do or how we would do it, I knew only that we had to do it. Once again, I had only the simple thought in my mind "whiskey, how cool is that?" I knew from an analytical chemistry standpoint that GC-MS could provide the information we wanted, so we were guaranteed success. How? We could figure that out.

Cleveland Whiskey: A Technology Company That Just So Happens To Make Whiskey

What I learned that evening during my first tour of Cleveland Whiskey was that Cleveland Whiskey is primarily a technology company. It just so happens that the product this technology company manufactures is whiskey. Recall that, traditionally, distilled spirit is aged in charred oak barrels for up to 10 years or more to flavor the whiskey with compounds that leach into the spirit from the barrel. Cleveland Whiskey founder Tom Lix had developed an innovative technology that accelerates the aging process of whiskey from a few years to a few days. Lix calls his accelerated aging process rapid "pressure aging." Although the specific details of the patented pressure aging process are proprietary, the procedure essentially involves placing the new spirit in a stainless steel vessel with pieces of charred wood of a controlled surface area, sealing the vessel, and subjecting the head space above the liquid to a precisely defined cycling of pressure (*2*). The pressure forces the alcohol into the wood, extracting compounds from the wood that naturally flavor the whiskey. Lix cleverly calls the pressure vessel used to mature whiskey, shown in Figure 1, R2D2. This revolutionary technology accelerates the aging/maturing process of whiskey from a few years to a few days, drastically reducing the production cycle timeframe.

Traditionally, oak barrels have been used for aging whiskey because oak is a hard and durable wood that enables the barrel to maintain its integrity over the long aging period. Owing to the pressure-aging technology that Cleveland Whiskey has developed, barrels have become antiquated and Cleveland Whiskey is therefore no longer confined to aging solely with oak wood. Subsequently, Cleveland Whiskey has proceeded to mature distillate with other varieties of wood generating new experimental flavors of whiskey that are unprecedented, including black cherry, apple, hickory, sugar maple, and honey locust. But again, I get ahead of the story.

Figure 1. The pressure vessel, R2D2 (back left), used for accelerated pressure aging of bourbon at Cleveland Whiskey. (Photograph courtesy of Cleveland Whiskey.)

As Cleveland Whiskey was founded on the basis of an innovative new technology for manufacturing whiskey, Cleveland Whiskey is essentially a technology company. It is therefore no surprise that a chance encounter between the founder and CEO of a technology company and a chemistry professor would quickly result in the establishment of a cooperative research project using GC-MS to study the flavor compounds in bourbons produced using this new technology. While the encounter happened by chance, the ensuing inception of the program proceeded organically and was therefore only pseudo-accidental.

If we were going to do this, I would first need to organize some infrastructure on our campus for doing the work, and I needed to act immediately.

Initiation of a Student Research Group

I had arrived on the campus of Lorain County Community College only a year earlier as a new faculty member. Upon arrival, I was immediately enthralled to find two faculty members in the Department of Biology, Professor Harry Kestler and Professor Kathy Durham, managing quite substantial research groups. Their groups were based on the graduate school model, with well-equipped laboratories and students performing substantial project-based independent research on an ongoing and self-perpetuating basis. It is challenging enough for four-year colleges to offer substantial research experiences to advanced undergraduate students. Yet here, at a two-year college where the challenge is even more substantial because students are only on campus for their freshman and sophomore years, I saw flourishing programs where students were performing not just a single isolated independent studies project, but ongoing research within the structure of a research group working on cooperative topics. I never would have dreamed it possible: graduate style research groups at a community college! But when I saw it, I knew that I wanted to do the same thing. Now I had discovered my topic: the "science of the flavor of whiskey."

First Students

I quickly recruited two outstanding students from my lecture classes to work on the project, Aubrie Thompson and Chris Kazee, Figure 2.

Figure 2. Spring 2015 Research Group: Aubrie Thompson (foreground, right) and Chris Kazee (background, left). (Photograph by Ronald Jantz.)

As stated, whiskey is a complex mixture of a large number of flavor compounds (up to hundreds), all present in very low quantities (circa ppm), and therefore the flavor industry recognizes GC-MS as a routine technique for application to study these complex mixtures (*3*, *4*). Techniques for improving chromatographic resolution (*5*), optimizing separation (*6*), and quantifying results (*7*) are commonplace for GC-MS studies of flavor compounds in spirits.

Our cooperative project with Cleveland Whiskey was initiated by using GC-MS to identify and quantify the volatile flavor compounds present in varieties of rapid pressure-aged bourbons alongside traditionally aged bourbons. The objective was to compare the flavor profiles of bourbons produced by rapid pressure aging to bourbons which had been traditionally aged. In our first semester, the initial group of students learned how to collect these data for these samples and began to generate results. We observed that bourbons produced by rapid pressure aging were nearly identical to bourbons which had been traditionally aged, the only observable difference in volatile compounds being that bourbons produced by rapid pressure aging included the addition of steric acid, a compound with little to no associated flavor (*8*). This was a welcome result for Cleveland Whiskey, as the technology company now had scientific data to assert that their rapid pressure-aging process yields bourbons equivalent in flavor to traditionally aged bourbons.

Continuing the Beginning

At the conclusion of this first semester, Chris Kazee transferred to a four year school, as many community college students do, when he had maxed out on the number of credits which the four year college would accept for transfer. While this left me with only one student, it did leave me with an experienced and extraordinarily talented student, Aubrie Thompson, Figure 3.

Figure 3. Fall 2015 Research Group; Aubrie Thompson. (Photograph by Ronald Jantz.)

I knew in my heart that we were on to something with this topic, "the science of the flavor of whiskey." I knew that it had to be developed if we wanted to grow to the next level. Frankly, I knew that we needed money.

I had an open and candid conversation with Aubrie, asking her if she would be willing to continue her work on the project into a second semester. I knew that Aubrie was capable of working independently. The plan was that I would devote my time to drafting as many proposals as I could find appropriate grant calls, while Aubrie would refine our lab technique to improve the signal-to-noise ratio of our data.

Aubrie successful refined our technique, improving the signal-to-noise ratio and thereby yielding higher-quality chromatograms. She also began to analyze some new whiskey samples of experimental new flavors from Cleveland Whiskey. Before Aubrie graduated that semester, I was left with a laboratory protocol which future students could use to collect the data that we wished to have for our whiskey samples.

External Confirmation of the Validity of Our Endeavors

While Aubrie worked diligently analyzing whiskey, I spent every free moment writing grant proposals. I ultimately submitted three.

The first was to the American Chemical Society (ACS) Collaborative Opportunities Grants program. Receipt of this grant award provided a sort of motivational shot-in-the-arm, a validation that someone from the outside saw our endeavors as meaningful and worthwhile (*9*). This was the encouragement we needed to ramp-up our efforts and bring the program that we were developing to the next level.

A second grant award was received that same semester from the ACS Two-Year College Faculty/Student Travel Grant Awards program, enabling travel to our first national meeting to present our work on the project. Finally, a third grant award was received from the Lorain County Community College Center for Teaching Excellence, providing for the balance of our travel to the national ACS conference along with our impending travel to other local and national conferences.

I quickly understood that no one grant award covers 100% of any expense. I quickly understood how securing and putting together multiple sources of grant funding was essential.

The Program Blossoms

Upon Aubrie's graduation I was left with the highly unfavorable position of no experienced students. Nonetheless, with funding secured I was ready to grow the effort. Recruiting four new students to work on the project, we were now a research group, as opposed to one student, Figure 4. I recruited the four new students, Katie Nowlin, Valerie Gardner, Christopher Wright and Clayton Mastorovich, again from previous lecture classes. With a laboratory analysis method developed, funding secured, and students recruited, we were now well situated to accomplish some substantial science.

Figure 4. Spring 2016 Research Group: (front-back) Katie Nowlin, Valerie Gardner, Christopher Wright, Clayton Mastorovich, Coleen McFarland (Envantage, Inc.), Professor Regan Silvestri. (Photograph by Ronald Jantz.)

Exciting Results for Unprecedented Bourbon Flavors

Recall that traditionally, oak barrels have been used for aging whiskey because oak is a hard, durable wood that enables the barrel to maintain its integrity over the long aging period, thus minimizing evaporation of the volatile product. Again, owing to the pressure-aging technology that Cleveland Whiskey has developed, barrels have become antiquated and Cleveland Whiskey is therefore no longer confined to aging solely with oak wood. Subsequently, Cleveland Whiskey had proceeded to mature distillate with other varieties of wood generating new experimental flavors of whiskey that are completely original, unprecedented, and made possible only via the innovative technology of accelerated pressure aging. We had just been provided with samples of some experimental batches of unprecedented bourbon whiskey flavors including black cherry, apple, hickory, sugar maple, and honey locust. These unprecedented bourbons are naturally flavored with compounds extracted from the various unique woods used for maturing via pressure-aging. The technology of pressure-ageing essentially allows the capture of previously untapped natural flavors. Imagine how thrilled chemistry students at the college were to receive samples of experimental whiskey flavors that were not yet commercially available!

Essentially, routine straight injection GC-MS was used to identify and profile the distinct flavor compounds that were leached from the various woods into these uniquely flavored bourbon whiskies (*10*). It was found that in general, the various woods impart largely the same flavor compounds to the bourbons, however in relatively different quantities (*11*). Table 1 lists the peak assignments for the GC-MS chromatogram traces along with the corresponding flavors of the compounds thereby identified.

Essentially, the flavor profiles of each of these uniquely flavored bourbons were determined, allowing comparisons between the different flavors of wood. For example, Figure 5 shows an overlay of the chromatogram trace for a bourbon finished with black cherry wood as compared to one finished with American oak. Accordingly, it can be seen that cherry bourbon has more ethyl octanoate, a compound known to impart a sweet fruity flavor, than does traditional oak flavored bourbon. Further, it can be seen that cherry bourbon has less phenethyl alcohol, a compound known to impart a floral and bready flavor, than does traditional oak bourbon (*12*).

The comparison of cherry to oak aged bourbon is demonstrated herein as only one example of the types of data that can be produced via application of this method. Further details of the data and interpretations (*11*) are omitted, as the scientific results are not really the story that we wish to tell here. Ultimately, comparisons between the relative quantities of the various flavor compounds were carried out in detail for all of the experimental bourbons aged with these various unprecedented woods. These unprecedented flavors of bourbon, black cherry, apple, hickory, sugar maple, and honey locust, have since been commercialized. Essentially, the routine technique of straight-injection GC-MS was applied to yield straightforward and valuable results: analytical descriptions of the flavor profiles of new flavors of bourbons.

Table 1. GC-MS Peak Assignments for Unprecedented Flavors of Bourbon

Retention Time (minutes)	*Compound*	*Flavor*
2.05	furanone	sweet, fruity, caramellic, brown maple note
2.10	branched butanol/ pentanol	sweet, fruity, fusel oil, apricot, banana, apple, wine
2.20	branched butanol/ pentanol	sweet, fruity, fusel oil, apricot, banana, apple, wine
2.30	glycolate	fruity, green, pineapple, fusel
2.40	phenyl acetaldehyde	honey, sweet, floral, chocolate and cocoa, with a spicy nuance
2.75	furfural	woody, sweet, bready, nutty, caramellic with a burnt astringent nuance
2.90	isoamyl acetate	sweet, banana, fruity with a ripe estery nuance
3.10	glycolic acid	very mild buttery
3.30	benzoic acid	faint balsamic
3.40	5-methylfurfural	sweet, brown, caramellic, grain, maple-like
3.50	ethyl hexanoate	sweet, pineapple, fruity, waxy, banana
4.25	phenethyl alcohol	floral, sweet, rosey and bready
4.55	ethyl octanoate	sweet, waxy, fruity, pineapple, creamy, fatty, mushroom, cognac notes
4.70	decanoic acid	fatty citrus, waxy fruit
4.80	stearic acid	none
4.85	decanol	waxy, fatty, tart, perfumistic, floral orange, sweet clean watery
4.90	phenethyl acetate	sweet, honey, floral, rosy, slight green nectar, fruity body
5.05	ethyl cyclohexane propionate	fruity, sweet, pineapple, peach, pear, honey, caramallic, maple and cocoa
5.10	benzoic acid	faint balsamic
5.25	furanone	sweet, fruity, caramellic, brown maple note

Peak assignments for GC-MS chromatogram traces of bourbons aged with unprecedented woods and corresponding flavors of the compounds.

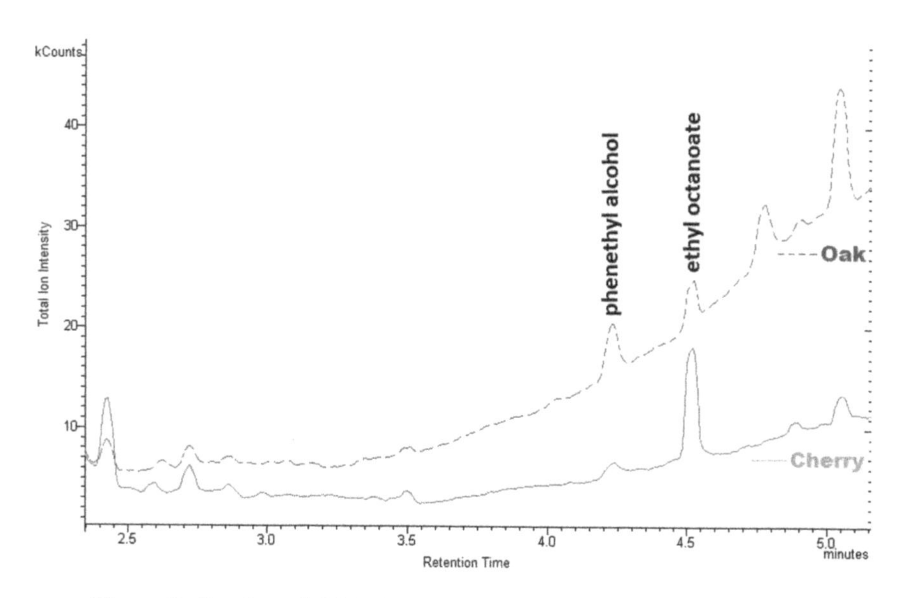

Figure 5. Overlay of GC-MS traces of cherry and oak aged bourbons.

On the Road

With data and results generated, what happened next was a whirlwind of activity of the students presenting the results of our work at various local and national conferences. Thanks to funding from multiple sources we were able to support student travel to nine conferences in one semester, a pace which we would ultimately maintain. The adventure of this aggressive travel schedule solidified my opinion on two things: (1) the value of the experience for students of presenting in public, and (2) the need to put together funding from multiple sources to support student travel.

In the News

We thought it was a great story: college students excited about researching the science of the flavor of whiskey. Non-chemistry students were willingly volunteering to participate in an optional program in chemistry. The marketing department at our college prepared a press release, which had already made it into multiple publications in our first semester (*13*, *14*). However, the media attention was about to grow sizably.

Receipt of the ACS Collaborative Opportunities Grant quickly brought national attention to our program. First we were invited to publish on the ACS Undergraduate Student blog "Reactions (*10*)." Next, Chemical and Engineering News (C&EN) magazine prepared to print an article featuring the winners of the ACS Collaborative Opportunities Grant awards (*15*). In the process of this article being prepared, the editor of C&EN became aware of our work and we were promptly featured in the Newscripts column of C&EN: the holy-grail of media attention, the "what's cool in chemistry news" feature (*12*). The Newscripts publication was re-posted on a multitude of formats before the print copies of the magazine were even shipped. Our student Christopher Wright was recognized while standing in line at Starbucks for a coffee in the hotel lobby at the ACS national meeting in Washington, DC.

This was promptly followed by a feature article in inChemistry magazine, the ACS undergraduate student magazine (*16*), which was subsequently followed by numerous additional features in the media (*17*, *18*).

Continuing Forward and Upward

Our aggressive travel schedule and the publicity which we had been fortunate to garner resulted in significant local and national attention to our program. The word "whiskey" makes for a great headline but unfortunately some of the off-the-cuff feedback seemed to downplay the science to the point of suggesting that it was all just-for-fun. I felt the need to prove that, while the samples we were analyzing were whiskey and it was fun, we were indeed performing concrete and significant scientific work. My idea was to demonstrate that I could perform this work with students who were minors and not yet of age to be legal adults. High school students can do chemistry, right? At our community college we have two programs that attract high school students to study on our campus: College Credit Plus (CCP) and Early College High School (ECHS). While they are two distinct programs, both serve to attract students who are excelling at high school to the college campus to complete their high school educations earning college credits. Bottom line, these are exceptional students, so they are highly capable, and participating in independent research would not be troubling for them.

Figure 6. Fall 2016-Spring 2017 Research Group: (left-right) Selena Vazquez, Christopher Wright, Clayton Mastorovich, Professor Regan Silvestri, Daniel McKeighen, Katie Nowlin, Valerie Gardner, Heather Ketchum. (Photograph by Ronald Jantz.)

I recruited one CCP student, Heather Ketchum, and one ECHS student, Selena Vazquez, to join the group. All four of the students from the previous semester remained on campus and continued in the research program, so there was a lot of expertise which could be used to train the new students. In addition to the two new high school students, one new traditional college student, Daniel

McKeighen, also joined the group, Figure 6. We were now a research group modeled after the graduate school protocol: well-equipped, well-funded, with more senior experienced students who could train incoming new students.

So where would we turn our scientific focus to now?

Chinese Baijiu Spirit Flavored as American Whiskey

China and Japan are seen in general by American industries as enormous market opportunities. The American bourbon industry is certainly no exception, which is currently looking toward Asia with much anticipation for continued largescale growth (*19*).

Baijiu is a popular Chinese spirit distilled from fermented sorghum. While this clear liquor is considered strong in exotic flavors by the western palate, its flavor is considered a standard by the eastern palate. In an effort to modify the flavors of baijiu only slightly, samples of various baijiu spirits were matured with wood by Cleveland Whiskey. The motivation was to create a new product for the eastern market: a recognizable flavor (baijiu) with an American twist (wood).

Toward this goal, experimental samples were produced of various Chinese baijiu spirits flavored to taste more similar to American bourbon by aging with various woods. This was accomplished by processing Chinese baijiu liquors via the novel accelerated aging process of pressure aging to mature the spirit in 24-48 hours and impart wooden barrel flavors. By processing Chinese baijiu liquor via this innovative technology of rapid pressure aging, the clear spirit becomes colored and flavored with the wood in the short time of a few days.

The student research group, now well versed in applying GC-MS to characterize flavor profiles for new flavors of spirits, turned their attention to these newly created liquors: American-styled Chinese spirits.

Multiple different imported samples of commercially available baijiu liquors were used as starting material and each was matured by finishing with charred American oak, black cherry, and honey locust woods. Analogous to our previous work with oak aged bourbons, the complete peak assignments for the GC-MS chromatogram traces of these wood aged baijiu spirits were determined along with the corresponding flavors associated with the compounds thereby identified (*20*). While a wood matured Chinese baijiu may be complete terra incognita, some published results do exist for GC-MS peak assignments of flavor compounds in traditional neat baijiu which were used to validate our peak assignments (*21–23*).

Without listing the detailed chromatogram peak assignments and their corresponding flavors, ultimately it was observed that wood aged Chinese baijiu flavor is dominated by a series of unbranched aliphatic ethyl ester compounds (*20, 24, 25*). A complete and perfectly sequential series of unbranched ethyl esters was observed: from ethyl butanoate, through ethyl pentanoate, ethyl hexanoate, ethyl heptanoate, to ethyl octanoate. Further, it was observed that certain baijiu starting materials were more altered in flavor by the pressure-aging process than other starting materials, indicating that different baijiu starting materials respond to the pressure-aging process differently (*20, 24, 25*).

Overall, baijiu flavor is dominated by ethyl hexanoate. Further, a higher concentration of ethyl hexanoate is observed in charred oak aged baijiu relative to traditional neat baijiu. The higher concentration of ethyl hexanoate in charred oak aged baijiu can be inferred as to impart a sweet and fruity nuance to the spirit relative to the traditional unaged neat baijiu. Whether this is desirable may be debated by whiskey enthusiasts endlessly in a taste test over a glass neat, a glass on ice, a glass with a splash of room temperature water, or however one believes that it should be tasted.

On the Road… Again

New data in hand, we continued our aggressive travel schedule and in academic year 2016-2017 students presented their work at 13 conferences, some local and some national. As had now grown to be standard protocol, all of this travel was funded thanks to support from external grant funding. We had not (yet) asked the college or Cleveland Whiskey to support our efforts financially. I was, now as a matter of practice, submitting on average three grant applications every semester for this project alone. The grant awards have all been quite modest, but we are a community college and we are accustomed to accomplishing things on a shoe-string budget. It is our modus operandi.

The student presentations promoted awareness of our work which brought more press. Most notably, I had asked Heather Ketchum, one of the new high school students, to present her work at the annual meeting of the Ohio Academy of Sciences (*24*). The Ohio Academy of Sciences annual meeting invites college students to submit abstracts for oral podium presentations whereas high school students are limited to submitting abstracts for poster presentations. I reasoned that Heather, while a senior at Brookside High School, was also a student at Lorain County Community College (via the CCP program), and we therefore submitted an abstract for her to give an oral presentation. Heather became the first high school student in the history of this conference to make an oral podium presentation; this ceiling exceeded, not by a prestigious liberal arts college or a large research university but by a community college, of course once again garnered attractive press.

Where to Next?

The start of the new academic year meant recruitment of new students into the group. By now, it seemed that everyone on campus was aware of the "science of the flavor of whiskey" project. In fall of 2017 the size of the research group grew to fifteen students, Figure 7, and in the spring of 2018 the group has grown to twenty students.

The group had grown larger than I ever envisioned, larger than I ever intended, and frankly larger than I could handle. It seemed the time had come for me to make a decision when I realized that the group was larger than I could handle alone. The idea was to hire an adjunct faculty member to be a research manager who would work directly with the students. The adjunct faculty salary for the research manager would be funded by student registrations for independent research, so the model was sustainable. As long as we had students, we could continue to employ a research manager. And I was easily able to persuade the Dean of Science and Mathematics to allow me to hire a research manager when I demonstrated how the funds for the salary would not be a deficit to the Division of Science and Mathematics budget, but would be funded by student registrations for independent research.

Figure 7. Fall 2017 Research Group: (front row left-right) Helen He Weizhuan, Lillian Williams, Selena Vazquez, Katie Nowlin, Valerie Gardner, Alyssa Adkins, S Jay McIntyre, Zoe Cornwall, (back row left-right) Clayton Mastorovich, Josh Niemiec, Mystal Jackson, Dr. Leanna Ergin, Professor Regan Silvestri, Ryan Reffner, Daniel McKeighen, (not pictured) Christopher Wright, Edward Tanos, Professor Jeff Zeager. (Photograph by Ronald Jantz.)

My role would now be relegated to writing proposals, securing funding, identifying presentation opportunities for students, giving presentations myself, identifying publication opportunities, and recruiting new students. In the lab, the students would now be guided through their research by Dr. Leanna Ergin, a meticulous scientist with great attention to analytical detail and a natural talent for working with students.

I now believe that the decision to hire a research manager has been the single most significant factor in defining the future history of our program. My efforts were now more focused, and more significantly the students were now getting a lot more attention. It is not surprising that we immediately began to accomplish a lot more science.

Research manager in place to guide us, where would we turn our scientific focus to now?

Infusion of Bourbon into Barbeque Smoker Chips

The wood used during the accelerated pressure-aging process becomes drenched in bourbon due to the associated pressure. Cleveland Whiskey astutely realized that this wood should not be a waste stream, but a byproduct which could itself be marketed as a product. Hence the inception of a new product line: bourbon infused barbeque smoker bricks. Demand for this new product quickly grew, and Cleveland Whiskey was faced with the prospect of increasing production.

We were asked by Cleveland Whiskey to optimize the process variables for the infusion of bourbon into barbeque smoker chips. We set up an experimental design to determine the process conditions that would lead to maximum infusion of bourbon into the bricks, and furthermore, how to reach maximum infusion of bourbon in the timeliest fashion possible.

The results of this work being of value to Cleveland Whiskey, we were offered, without asking, funding support by Cleveland Whiskey. How appropriate. Two years ago we received our first funding, a grant from the ACS via their Collaborative Opportunities Grant program. The purpose of the Collaborative Opportunities Grant program is to establish long-term relationships between community colleges and various community partners such as industry. The activity supported by this grant had allowed us the opportunity to demonstrate our value to Cleveland Whiskey, our industry partner on that grant.

We proceeded to successfully optimize the process variable parameters for the infusion of bourbon into barbeque smoker bricks. Regrettably, we cannot share the scientific results of this work as the information is confidential, work-for-hire performed for Cleveland Whiskey.

Looking Back with Hindsight Vision

Impact on Students

Of the students thus far involved in the research program, only two are studying to be chemists and one a science teacher. Certainly we can presume it highly likely that none of the students will become flavor scientists. So what value does participation in the program bring these students?

The overwhelming majority of the students who have thus far participated in the program are health science students. This experience has given them a genuine understanding of the way that analytical data is generated and processed. I envision a scenario in their future careers where they will say: "We need information. Let's send a sample to the lab. If we talk to the people in the lab, they can advise on what would be the appropriate test to run." This experience has left the students with a predisposition that analytical data can be interpreted to provide useful information.

As a two-year community college, our students transfer to four-year colleges to complete bachelor's degrees. Three of the students currently in the research group have already transferred to four-year colleges and, despite being transferred from our community college, they have continued to return to our campus on Friday afternoons to participate in the whiskey research program. This demonstrates how connected the students have grown to our campus community via this program. Clearly the experience of being a part of this research group has instilled the students with a sense of belonging. Even as commuter students, they are nonetheless engaged on our community college campus thanks to this program. Student engagement? Students, the majority of whom are not chemistry majors, are willing to spend their free-time on Friday afternoons in a chemistry lab for no reason other than they want to.

Further, every student who has been through the program has presented their work publicly at a local or national conference. The true value for the students of participation in the program has been the opportunity for them to speak publicly. They learn how to convey highly technical information to a non-expert audience, and how to do so succinctly. Arguably, this is invaluable life experience.

Lessons Learned

The initial chance encounter with Cleveland Whiskey led to the initiation of a collaboration to establish this program. This chance encounter may on the surface seem accidental, but we would argue that it was only pseudo-accidental. Looking back on the circumstance of the initial encounter, this chance meeting happened (1) because I invest a great deal of time staying in contact with former students (one of whom completed an internship at Cleveland Whiskey), and (2) because I am active in local scientific societies (and organized a meeting at Cleveland Whiskey.)

In hindsight, an industrial partner has been essential for developing a research program which is practical and applied. Students find the work concrete and easily relatable, as they are generating practical and useful information. Whereas academic research is often seen as conceptually abstract, students find this work concrete and conceptually straightforward. As a result, students are encouraged when they are able to quickly contribute to the work. In short, the analytical technique is well established, it can be applied to unique samples, and as such yields practical and useful information.

As students find the work concrete and easily relatable, we are able to recruit students into the program in their first semester on campus. Specifically, students in General Chemistry 1 do not need a detailed understanding of the theoretical basis of GC-MS to understand how the instrument can be applied to provide useful information on a practical topic. The lesson learned is that undergraduate research need not and should not be reserved for upperclassmen students simply because they have advanced to upper-level coursework.

As students are presumably to conduct independent research "independently", faculty efforts can focus dominantly on coaching students in public presentation skills. Indeed, the experience of presenting such highly technical information to a non-expert audience, and conveying the information succinctly, has been the most valuable experience gained by the students who have participated in the program.

Also, faculty efforts can focus on identifying opportunities for students to present publicly and securing funding to support student travel. It is rare for a single grant opportunity to fully support a project. Most grants offer partial support of a project. This is especially true when it comes to travel lines in the budget. As such, multiple sources of funding, each which partially support, need to be put together. The job of the faculty, as I see it, is to put everything together. I believe that the key to our on-going success with this project has been the securing of multiple sources of funding.

Further, our success rate with grant proposals has been extraordinary, with 92% of our applications receiving funding. I believe that this is because we have chosen to submit to calls for very modest grants. Grant calls with lower dollar amount limits typically have higher award rates, and further have simpler application and reporting procedures. Despite that each of the individual grant awards we have received is modest, we have nonetheless put together something that is quite substantial when it is all added together.

This program was established based on the simple thought: "whiskey…cool, let's do that." Nothing was planned more than one (or two) steps in advance. Certainly we never envisioned that the program would grow to become what it has. The program grew one step at a time by seeing an opportunity and taking action to secure it. Simply, a good work ethic, together with organized workmanship and the involvement of talented and genuine people, has advanced this program to the level of success that it has achieved.

Finally, we believe that the large amount of success which this program has enjoyed is due predominately to one word: whiskey. For example, I purposefully used the word whiskey in the title of this paper to get you to read it, and here you are. Certainly, the word whiskey has put the project into media headlines, which is excellent public relations for the college. Further, the choice of a topic (whiskey) is paramount for inspiring student engagement, especially in an optional undergraduate research program. Owing to the whiskey connection, students are enthusiastic to become involved and they enjoy the experience.

Summary

The innovative production technology of rapid pressure-aging allows endless possibilities to innovate new flavors of whiskey. Experimental batches of pistachio bourbon and mesquite bourbon have already been produced, to name only a few. Essentially, the technology opens possibilities for endless untapped flavors from unprecedented woods, where creativity is now the only limit to innovation. This work can continue into the future in unlimited different directions bound only by creativity.

A vibrant independent student research program for undergraduates has been created and is blossoming, with students working independently on various aspects of a topic as part of a larger collaborative research group. A research group based on the graduate school model has been established at a two-year community college, with a well-equipped and well-funded laboratory, and students performing substantial project based independent research on an ongoing and self-perpetuating basis (*16*). While the quantity and depth of the research does not compare with graduate work, the model that the work is conducted under is identical thus preparing students for what they will encounter in their futures. Students are performing not just a single isolated independent studies project, but ongoing research within the structure of a research group with colleagues working on cooperative topics. A graduate style research group at a community college!

A graduate modeled research group has been established at a community college and is flourishing. Students are enthusiastic to become involved in the "science of the flavor of whiskey" project, which has quite the buzz on campus.

This is the story of how a successful undergraduate student research program can be put together: from inception, to involvement, to funding and sustainability.

Acknowledgments

Grateful thanks is acknowledged for financial support from: the American Chemical Society Collaborative Opportunities Grants program, the American Chemical Society Two-Year College Faculty/Student Travel Grant Awards program, the Lorain County Community College Foundation Campus Grants program, the Lorain County Community College Center for Teaching Excellence, the NASA Ohio Space Grant Consortium, the Ohio Academy of Sciences, Ohio Means Internships and Co-Ops, Choose Ohio First, and also thanks to corporate sponsorship from Cleveland Whiskey.

References

1. Gill, V. A Whisky Tour. *Chemistry World*, December 2008, pp 40−44.
2. Lix, T. S. *Whiskey Making Method.* U.S. Patent 0149423A1, 2013.
3. Scott, D. *Aroma Study of Potable Spirits*; Application Note; PerkinElmer: Shelton, CT, 2013.
4. *Analyzing Alcoholic Beverages by Gas Chromatography*; Technical Guide, Lit. Cat. No. 59462; Restek: Bellefonte, PA, 2002.
5. *Improved Resolution of Alcoholic Beverage Components by Packed Column GC*; Bulletin 790C; Supelco: Bellefonte, PA, 1996.
6. Wittrig, R. E.; Macnamara, K.,; Mosesman, N. *Optimizing the Capillary GC Separation of Acids, Esters, and Other Flavor Components in Distilled Liquor Products*; Technical Guide; Restek: Bellefonte, PA, 2002.
7. *Qualitative Comparison of Whisky Samples Using Fast GC/TOFMS*; Corporation Form No. 203-821-200; LECO Corp.: Saint Joseph, MI, 2008.
8. Thompson, A.; Kazee, C.; Silvestri, R. GC-MS Analysis of the Flavor of Bourbon Whiskey Produced by a Novel Accelerated Aging Process. In *Proceedings of the Society for Applied Spectroscopy 59th Annual May Conference*, May 20, 2015, p 67.
9. *Independent Student Research: Flavor of Whiskey*. American Chemical Society. https://www.acs.org/content/acs/en/funding-and-awards/grants/collaborative-opportunities-grants/independent-student-research--flavor-of-whiskey.html (accessed January 31, 2018).
10. Silvestri, R. *Chemistry of Whiskey Flavor*, American Chemical Society Undergraduate Blog *Reactions*, April 6, 2016. https://acsundergrad.wordpress.com/2016/04/06/chemistry-of-whiskey-flavor/ (accessed January 31, 2018).
11. Nowlin, K.; Gardner, V.; Mastorovich, C.; Wright, C.; Silvestri, R. GC-MS Analysis of Unprecedented Whiskey Flavors Created by a Novel Aging Process. In *Proceedings of the Society for Applied Spectroscopy 60th Annual May Conference*, May 18, 2016, p 63.
12. Widener, A. Imparting Whiskey Wisdom. *Chem. Eng. News* **2016**, *94*, 40.

13. Carrasquillo, K. The Science of Whiskey. *Career Focus*, Fall 2015; p 3.
14. Carrasquillo, K. The Science of Whiskey. *Update*, Summer/Fall 2015; p 2.
15. Wang, L. Collaborative Opportunities Grants Available. *Chem. Eng. News* **2016**, *94*, 35.
16. Silvestri, R. Distilling Lessons from a Whiskey Research Program. *inChemistry* **2016**November, 12–13.
17. Local Students in the National Spotlight. *Isotopics*, October 2016; p 2.
18. Institutional Member Spotlight: Lorain County Community College Cleveland Whiskey Project team. *Ohio Academy of Science Newsletter*, October 26, 2017.
19. Horn, J. Why Japan Is the Best Country for Whisky Right Now. *Playboy*, 2016.
20. Silvestri, R. In *Best Practices for Supporting and Expanding Undergraduate Research in Chemistry*; Gourley, B.; Jones, R., Ed.; ACS Symposium Series 1275; American Chemical Society: Washington, DC, 2018.
21. Zheng, Y.; Sun, B.; Zhao, M.; Zheng, F.; Huang, M.; Sun, J.; Sun, X.; Li, H. Characterization of the Key Odorants in Chinese Zhima Aroma-type Baijiu by Gas Chromatography–Olfactometry, Quantitative Measurements, Aroma Recombination, and Omission Studies. *J. Agric. Food Chem.* **2016**, *64*, 5367–5374.
22. Xu, H.; Xu, X.; Tao, Y.; Yuan, F.; Gao, Y. Optimization by Response Surface Methodology of Supercritical Carbon Dioxide Extraction of Flavour Compounds from Chinese Liquor Vinasse. *Flavour Fragr. J.* **2015**, *30*, 275–281.
23. Wu, J.; Zheng, Y.; Sun, B.; Sun, X.; Sun, J.; Zheng, F.; Huang, M. The Occurrence of Propyl Lactate in Chinese Baijius (Chinese Liquors) Detected by Direct Injection Coupled with Gas Chromatography-Mass Spectrometry. *Molecules* **2015**, *20*, 19002–19013.
24. Ketchum, H.; Silvestri, R. GC-MS Analysis of Chinese Baijiu Liquor Flavored as American Bourbon Whiskey. *Ohio Journal of Science* **2017**, *117*, 6.
25. Gardner, V.; Silvestri, R. GC-MS Analysis of Unprecedented Whiskey Flavors Including Chinese Baijiu Flavored as American Bourbon. *Proceedings of the Society for Applied Spectroscopy 61st Annual May Conference*, May 24, 2017, p 52.

NSF Funding Programs

Chapter 12

What To Know Before You Write Your First NSF Proposal

Thomas B. Higgins*

Department of Chemistry, Harold Washington College, 30 E. Lake St., Chicago, Illinois 60601, United States
***E-mail: tbhiggins@ccc.edu.**

From 2015-2017, I had the opportunity to serve as a rotating Program Officer (PO) in the Division of Undergraduate Education (DUE) at the National Science Foundation (NSF). This experience not only allowed me to serve my country and my scholarly community, but it also gave me a richer perspective on the process of obtaining NSF funding to support students. This paper will attempt to provide general advice for the two-year college chemistry community on the NSF funding process, with an emphasis on information for the first-time principal investigator (PI). This advice will be most relevant to NSF programs in DUE, although hopefully it will be helpful with programs in other NSF divisions as well. This paper is also meant to complement rather than repeat the good advice provided by other members of the chemistry community in other ACS publications (*1*).

Introduction

The NSF has an interesting history. It was created by Congress in 1950 during the aftermath of the Second World War, and one of the drivers for its creation was the influential report "Science: The Endless Frontier" by Vannevar Bush (*2*, *3*). This report highlighted the importance of both fundamental and applied science and scientific research for ensuring the security and wealth of the nation and the free world. It formed the basis of the broad charge given to NSF: "…to promote the progress of science; to advance the national health, prosperity, and welfare; to secure the national defense (*4*)..." This charge informs all of NSF's programs and activities. An overview of NSF, its history, the work it does, and the range of projects it funds can be found on the agency website (*4*).

The NSF is a small federal agency and uses its funds strategically. NSF funding success rates average approximately 20% across all programs, although this varies based on proposal pressure, funding levels, and other factors. Therefore, NSF POs must ensure that the proposals they recommend will fund projects that have the greatest potential to benefit not only PIs, their students, and their home institutions; but also the national education and research communities. These factors have a practical consequence for success with NSF funding requests: most proposals are not funded on their first submission and require revision and resubmission before being recommended for an award. This means that tenacity and grit can be key attributes for successfully obtaining an NSF grant. Luckily, the NSF merit review process is set up to provide PIs with valuable feedback from both peer reviewers and NSF POs. Effective use of this feedback is another essential component for success obtaining an NSF grant.

Getting Started: Finding a Program

The pursuit of an NSF grant starts with a good idea and, for proposals to DUE, one that is centered on evidence-based and evidence-generating activities that have potential to meaningfully improve students' learning, education, and preparation for the STEM workforce. Next, matching that idea to a specific DUE program is critical. NSF organizes its funding priorities into programs that are described in program solicitations. A solicitation contains specific guidance on the types of activities and outcomes NSF supports and expects, and the information PIs must provide in a grant proposal. Program solicitations change periodically, so it is critical to use the most up-to-date solicitation. Never assume a past solicitation is still in effect. NSF uses a document numbering system where the first two digits indicate the fiscal year in which the document was issued. Since solicitations are usually updated every 2-3 years, this number is a good indicator of how current the solicitation is. For example, publication NSF 18-1 was issued in the 2018 fiscal year.

For members of the two-year college chemistry community, the three most relevant DUE programs are Advanced Technological Education (ATE), Improving Undergraduate STEM Education: Education and Human Resources (IUSE: EHR), and Scholarships in Science, Technology, Engineering, and Mathematics (S-STEM). In addition, the NSF Robert Noyce Teacher Scholarships program offers an additional financial incentive for two-year/four-year partnerships that meaningfully involve community college students. A full list of all NSF programs across all divisions can be found on the "Funding" page of the NSF website (*5*).

The ATE Program

The ATE program focuses on technician education for the high-technology fields, including ChemTech and BioTech, that drive the nation's economy. It also requires that two-year college faculty members assume clear leadership roles in any project and that there are strong, documented partnerships with business and industry. A full synopsis of the program can be found on the ATE program homepage (*6*). This also includes a link to the program solicitation, a list of POs associated with the program, and a link to recently funded awards with abstracts. The current program solicitation as of the publication of this paper is NSF 17-568 and it should be in effect until 3 October 2019. After that date, the solicitation will be updated.

One common misconception about ATE is that it supports any project at a two-year college. This is not the case. All ATE proposals must have, at their core, a strong workforce education component that prepares students for a STEM career that is in demand by industry. If an idea does not have a clear technician education component, it is better to look at other programs for funding.

IUSE: HER Program

The IUSE: EHR program has a broad scope and focuses on supporting effective undergraduate STEM education for both majors and non-majors. Most ideas that impact the undergraduate classroom and the teaching environment will fall within the scope of IUSE. A full synopsis of the program can be found on the IUSE: EHR program homepage (*7*). There is also a link to the program solicitation, a list of POs associated with the program, and a link to recently funded awards with abstracts. A set of frequently asked questions and recorded webinars are also available. The current program solicitation as of the publication of this paper is NSF 17-590 and this should be in effect until 30 September 2019. After that date, the solicitation will be updated.

In December of 2017, IUSE unveiled a new program targeting Hispanic-Serving Institutions (*8*). This solicitation (NSF 18-524) was only for a single round of funding, but that does not mean it will go away. Since many community colleges are also HSIs, this is a good program to keep an eye on.

One common misconception about the IUSE program is that it is an incremental continuation of the old Transforming Undergraduate Education in STEM (TUES) and Course, Curriculum, and Laboratory Improvement (CCLI) programs. This is not the case. IUSE has a much stronger emphasis on knowledge-generation and this must be a cornerstone of the intellectual merit of a project (see below).

S-STEM Program

The S-STEM program focuses on providing scholarships for academically talented, low-income students with demonstrated financial need. And, because scholarships alone are not enough to ensure student success, it requires cohorting and mentoring activities that support students' academic success and degree completion. A full synopsis of the program can be found on the S-STEM program homepage (*9*). There is also a link to the program solicitation, a list of POs associated with the program, and a link to recently funded awards with abstracts. The current program solicitation as of the publication of this paper is NSF 17-527 and does not have a listed expiration date. That does not mean it will not change, however, so always check for a more current solicitation.

Noyce Program

The Noyce program supports creative proposals that recruit and prepare talented STEM majors and professionals to become highly effective K-12 STEM teachers. Community colleges play an important role in the education of future teachers and the Noyce program offers collaboration incentives in form of additional funding for projects that meaningfully engage community colleges and their students and faculty in project activities. These incentives are described in the "Award Information" section of the current Noyce solicitation, NSF 17-541. This solicitation does not have a listed expiration date, but that does not mean it will not change.

Understanding the Merit Review Process

The NSF merit review principles and criteria are the standards used to judge all proposals submitted to the NSF. The most up-to-date articulation of these standards can be found in the NSF Proposal Award Policy and Procedures Guide, otherwise known as the PAPPG (*10*). The PAPPG is usually updated each January, so always check the NSF website to find the most up to date version.

The PAPPG is a reference document, not a paper, so do not try to read it front to end. Instead, become familiar with the table of contents so you can find the information you need when you need it. Three critical parts of the PAPPG that all PIs should be familiar with are the:

- **Introduction, Section C.** "Listing of Acronyms" (Everyone at the NSF uses TLAs--Three Letter Acronyms. This is the definitive guide.)
- **Part I, Chapter II.** "Proposal Preparation Instructions" (Chapter II.C.2.g on "Budget and Budget Justification" is especially important.)
- **Part I, Chapter III.** "NSF Proposal Processing and Review" (This contains a description of the NSF merit review process, which is how the merit of a proposal is judged by external reviewers and internal NSF staff.)

Understanding the relationship between the PAPPG and the program solicitation is sometimes difficult for novice PIs. One way to think about it is that the PAPPG provides general guidance for writing an NSF proposal to any NSF program. On the other hand, the program solicitation provides specific guidance on a focused area of interest. Normally, these two documents complement each other. In the rare instances when they do not agree, the specific program solicitation has priority over the PAPPG.

The Merit Review Process

The NSF's merit review process uses a combination of external and internal perspectives to judge the merit of proposals submitted by PIs, and understanding the review process is an important factor in being successful with NSF funding. NSF provides a rich overview of the merit review process on its website (*11*), which also contains answers to frequently asked questions.

The merit review process consists of three distinct phases:

- Phase I: Proposal Preparation and Submission
- Phase II: Proposal Review and Processing
- Phase III: Award Processing

Phase I and Phase II are addressed in this paper. Phase III is not, as it is internal to the NSF. When a PI gets an award, he or she will become familiar with the steps of Phase III.

The merit review process is organized around two merit review criteria, which are critical to writing a successful proposal: intellectual merit (IM) and broader impacts (BI). Both criteria must be addressed in any proposal submission and both criteria require explicit statements in both the project summary and the 15-page project narrative. The requirement of an explicit IM statement in the project narrative is new with the 2018 revision of the PAPPG (NSF 18-1).

Intellectual merit is defined as "the potential to advance knowledge", so knowledge-generation must be a central part of any proposal submission. The IM component is sometimes referred to as "research", which is appropriate when one takes a big-picture view of what constitutes research, especially in an educational context. Not all research has to be oriented to the discovery of fundamentally new knowledge and processes.

To foster a deeper understanding of research in an educational context, the NSF and the US Department of Education developed the "Common Guidelines for Education Research and Development" (NSF 13-126) in 2013 (*12*). This document describes six levels of knowledge development, ranging from the generation of fundamental new and undiscovered knowledge to the implementation and deliberate study of interventions that are known to work in ideal settings but need to be understood in real settings such as the community college classroom. This latter type of research falls into the categories of efficacy, effectiveness, and/or scale-up. Anyone considering a submission to a

DUE program should read the Common Guidelines and describe their project's knowledge generation plans in the appropriate context.

Broader impacts are "the potential to benefit society." Often times, novice PIs either understate their BIs by not thinking beyond their own institution, or wildly overstate the significance of their ideas and over-promise what a project will do. The key to a meaningful BI statement is to think outside of your institution and be realistic about who can benefit from your work. One way to do this is to leverage relationships and collaborations outside of your institution, such as your membership in ACS and its divisions and local sections. To guide PIs' BI development, the NSF Office of Integrative Activities has published the report *Perspectives on Broader Impacts* (*13*).

In addition to the two merit review criteria, prior work and project evaluation are essential components of a successful proposal. Prior work can mean many things. The most important component is the results of prior NSF-funding, if any of the PI or the coPIs have had NSF funding in the past five years. In a prior work statement, it is important to show that a PI (1) meets his or her IM and BI goals, and (2) effectively shares his or her work with others.

Other important information related to prior work is institutional data that contextualizes what can be learned and why it is of interest, from both an IM and a BI perspective. Including a thorough literature review is also critical, so it is clear that a project is adding to the general knowledge base rather than reinventing the wheel. This must be adequately supported with citations from the relevant literature and other sources. All together, this discussion may consume 2-5 pages of a 15-page proposal narrative. When doing the background work for a project, use the NSF "Award Search" page to find colleagues working in similar areas (*14*). Each NSF-funded award has a publicly available abstract with contact information for the PI and a list of publications that have resulted from a project. Many PIs funded by DUE are happy to share their work, as this can open the door to future collaborations. New PIs should not be afraid to reach out to other people working in their area of interest.

Project evaluation is the systematic review of the implementation, progress, and impact of a project. This needs to be done by an external party. (Note that external means from the perspective of the project, not necessarily the institution. PIs and coPIs cannot evaluate their own work, but a qualified colleague in another department could.) A meaningful description of project evaluation should consume 1-2 pages of the project narrative. The evaluation plan needs to be clearly linked to the project's goals, objectives, outcomes, and activities. The credentials of the evaluator must be included as supplementary information, such as an NSF biosketch that highlights his or her expertise as an evaluator. If possible, engage a project evaluator while writing the proposal and have him or her contribute to the evaluation component of the narrative. Typically, evaluation should cost between 5-10% of the direct costs of a project's budget. For this, the evaluator should produce a formal evaluation report that the PI can include with his or her annual report to the NSF. For new PIs who are unfamiliar with project evaluation, the NSF-funded project *EvaluATE* (*15*) is a good resource, as is the report *The 2010 User-Friendly Handbook for Project Evaluation* (*16*).

What To Expect After Submission

After a proposal is submitted, it is reviewed by NSF POs to ensure it meets all the compliance requirements described in the PAPPG. If it passes this, and most do, it will be reviewed by a panel of peers. This procedure generates a set of reviews from each individual panelist and a single review from the panel as a whole. Next, the proposal is reviewed in depth by the PO. Based on all this information - panelist reviews, the panel discussion, and the PO's understanding of the current landscape and needs of the community - the PO makes a recommendation to either fund or decline a proposal. The ultimate decision to decline or fund a proposal is made by the DUE Division Director. This process can take up to six months from the submission date, so it is important to be patient.

Proposals that are declined are sent back to the PI along with the reviews and the PO's comments. This is valuable feedback, as most proposals must be submitted multiple times to result in an award. One official NSF estimate is 2.3 submissions per award (*17*). PIs should use this feedback to revise, update, and resubmit their proposals to the next round of competition. A key attribute to getting federal funding is persistence and revision. In the words of the wise Yoda, "The greatest teacher, failure is."

The Importance of Becoming a Reviewer

A great way to learn how to write meritorious proposals is to serve as a reviewer. Because NSF solicitations and the community landscape are constantly changing, this is good advice not only for beginning investigators but also for experienced investigators. In fact, some experienced investigators who write unsuccessful proposals do so because they are not writing to the most current solicitations. To become a reviewer is straightforward: simply email a PO to state your desire to review and attach a copy of your most recent CV. If you can talk to a PO in person, even better. POs are always happy to talk to prospective reviewers and PIs. More information on becoming a reviewer can be found on the NSF reviewer page (*18*).

Other resources to assist with writing meritorious proposals are also available. NSF holds both live and virtual "NSF Days" to stay connected with the community. A list of these events is published online and past events are archived on the web (*19*). Individual divisions and programs also have outreach events and post materials to guide PIs in writing high-quality proposals. These are usually found on the program solicitation home page, a division's activities page, or the "NSF News" site (*20*).

Finally, I found serving as a NSF PO was an incomparable professional development opportunity. I would highly recommend it for anyone interested. If you are interested, you can find details on this and other careers at NSF on the "Picture Yourself at NSF" page (*21*).

And remember: If you don't submit, the answer is always no!

Acknowledgments

This material is based upon work supported by the National Science Foundation under Grant No. 1655042. Any opinions, findings, and conclusions or recommendations expressed in this material are those of the author and do not necessarily reflect the views of the National Science Foundation.

References

1. *Nuts and Bolts of Chemical Education Research*; Bunce, D. M., Cole, R. S., Eds.; ACS Symposium Series 976; American Chemical Society: Washington, DC, 2008.
2. Zachary, G. P. *Endless Frontier: Vannevar Bush, Engineer of the American Century*; MIT Press: Cambridge, MA, 1999.
3. Pielke, R. In Retrospect: Science — The Endless Frontier. *Nature* **2010**, *466* (7309), 922–923.
4. *About NSF - Overview, NSF - National Science Foundation.* https://nsf.gov/about/ (accessed April 23, 2018).
5. *Funding: NSF - National Science Foundation.* https://nsf.gov/funding/index.jsp (accessed April 23, 2018).
6. *Advanced Technological Education, NSF - National Science Foundation.* https://nsf.gov/funding/pgm_summ.jsp?pims_id=5464 (accessed April 23, 2018).
7. *Improving Undergraduate STEM Education: Education and Human Resources, NSF - National Science Foundation.* https://nsf.gov/funding/pgm_summ.jsp?pims_id=505082 (accessed April 23, 2018).
8. *Improving Undergraduate STEM Education: Hispanic-Serving Institutions, NSF - National Science Foundation.* https://nsf.gov/funding/pgm_summ.jsp?pims_id=505512 (accessed April 23, 2018).
9. *NSF Scholarships in Science, Technology, Engineering, and Mathematics Program, NSF - National Science Foundation.* https://nsf.gov/funding/pgm_summ.jsp?pims_id=5257 (accessed April 23, 2018).
10. *PAPPG Introduction, NSF - National Science Foundation.* https://www.nsf.gov/pubs/policydocs/pappg18_1/index.jsp (accessed April 23, 2018).
11. *Merit Review, NSF - National Science Foundation.* https://nsf.gov/bfa/dias/policy/merit_review/ (accessed April 23, 2018).
12. *Common Guidelines for Education Research and Development, NSF - National Science Foundation.* https://www.nsf.gov/publications/pub_summ.jsp?ods_key=nsf13126 (accessed April 23, 2018).
13. *Perspectives on Broader Impacts, NSF - National Science Foundation.* https://www.nsf.gov/od/oia/publications/Broader_Impacts.pdf (accessed April 23, 2018).
14. *NSF Award Search: Simple Search, NSF - National Science Foundation.* https://nsf.gov/awardsearch/ (accessed April 23, 2018).
15. *EvaluATE.* http://www.evalu-ate.org/ (accessed April 23, 2018).

16. *The 2010 User-Friendly Handbook for Project Evaluation, EvaluATE.* http://www.evalu-ate.org/resources/doc-2010-nsfhandbook/ (accessed April 23, 2018).
17. *Merit Review: Facts, NSF - National Science Foundation.* https://nsf.gov/bfa/dias/policy/merit_review/facts.jsp#6 (accessed Apr 23, 2018).
18. *Merit Review: Why You Should Volunteer to Serve as an NSF Reviewer, NSF - National Science Foundation.* https://nsf.gov/bfa/dias/policy/merit_review/reviewer.jsp#1 (accessed April 23, 2018).
19. *Undergraduate Education (DUE) - Events, NSF - National Science Foundation.* https://www.nsf.gov/events/index.jsp?org=DUE (accessed April 23, 2018).
20. *News, NSF - National Science Foundation.* https://www.nsf.gov/news/ (accessed April 23, 2018).
21. *Picture Yourself at NSF, NSF - National Science Foundation.* https://www.nsf.gov/careers/ (accessed April 23, 2018).

Editors' Biographies

Laura J. Anna

Laura J. Anna is Professor of Chemistry at Montgomery College, a diverse, three-campus community college, in the DC metro region of Maryland. She has been teaching organic chemistry for over 20 years and is currently serving as Department Chair.

Thomas B. Higgins

Thomas B. Higgins is Professor of Chemistry at Harold Washington College, an urban community college where a majority of the students are from groups underrepresented in STEM. He has been teaching chemistry, astronomy, and physical sciences there for 21 years. In 2016, he was named a Fellow of the ACS, primarily for his work in broadening participation and his contributions to ACS governance.

Alycia Palmer

Alycia Palmer is Assistant Professor of Chemistry at Montgomery College where she teaches introductory and general chemistry. She is a strong supporter of Open Educational Resources and has worked to develop materials that are available to students at any institution for no cost.

Kalyn Shea Owens

Kalyn Owens is Professor of Chemistry at North Seattle College where she has been teaching general chemistry and STEM research courses for 15 years. She is passionate about designing innovative curriculum for early postsecondary STEM courses with emphasis on interdisciplinary curriculum, course-based research experiences and engaging students in social construction of new knowledge.

Indexes

Author Index

Subject Index

C

E

F

I

U